The Strategy Layer

Decisions That Drive Results

Steven Weaver

Producers Publishing

COPYRIGHT

Published by Producers Publishing

Disclaimer

This book provides frameworks and guidance for organizational decision-making and analytics strategy. The information is distributed on an "as is" basis, without warranty. While every precaution has been taken in the preparation of this work, neither the author nor the publisher shall have any liability to any person or entity with respect to any loss or damage caused or alleged to be caused directly or indirectly by the information contained in this book.

The frameworks and approaches presented should be adapted to your specific organizational context. Readers should seek appropriate professional advice for their unique situations. Results may vary based on organizational circumstances, resources, and implementation approaches.

Case Studies and Examples

Printed in the United States of America

CONTENTS

Introduction: From Strategy to Action

Every year, organizations invest billions in analytics. They buy the latest platforms, hire data scientists, and build impressive dashboards. Executives talk about being "data driven." Analysts work tirelessly to produce insights. Yet when it comes time to make important decisions, something fundamental is missing.

The data exists. The dashboards are polished. But the connection between all that information and the choices that actually shape organizational outcomes remains frustratingly weak.

If you work in analytics, you've felt this. You build comprehensive analyses that executives glance at briefly before deciding based on other factors. You create dashboards that get monitored but rarely drive action. You answer questions, but they're not quite the right questions.

If you're an operational leader or executive, you've experienced the other side. Despite massive investment in analytics, decisions still feel like they're made the same way they always were: through negotiation, institutional memory, and gut instinct. The data provides context, but it doesn't seem to make the hard choices any clearer.

This book solves that problem.

The Missing Layer

The gap between data and decisions isn't a technology problem. It's not about having better dashboards, more sophisticated algorithms, or cleaner data. Those things can help, but they're not the core issue.

The problem is structural. Organizations have strategy on one side and data on the other, but they've never built the explicit architecture that connects them. They've never designed decision making as a capability.

That's what the Strategy Layer provides: a systematic way to connect strategic goals to the specific decisions that drive those goals, and to connect those decisions to the metrics, data, and analytics that should inform them.

It's called the Strategy Layer because it sits between your strategic intent and your operational data, translating one into the other. It makes decision making visible, manageable, and continuously improvable.

What Makes This Book Different

This isn't another book about dashboards, data visualization, or analytics techniques. It's not about specific technologies or tools. It's about building the decision architecture your organization needs before any of those things can create real value.

It's practical. Every chapter includes step by step implementation guides, templates you can use immediately, and realistic examples from organizations like yours. You'll find exercises you can complete in 60 to 90 minutes that produce tangible results.

It's comprehensive. The book takes you from recognizing the problem through building complete decision support infrastructure to scaling it across your enterprise. You'll learn not just what to do, but how to do it and what to expect along the way.

It's proven. The frameworks and approaches in this book are drawn from real implementations across healthcare, manufacturing, financial services, technology, and other industries. They work in complex organizational environments with real constraints and politics.

Who This Book Is For

If you're an analyst or analytics manager, this book gives you the frameworks to shift from being a report factory to being a strategic partner. You'll learn how to insist on decision context, prioritize based on impact, and structure work that executives actually use.

If you're an operational leader, this book shows you how to get the decision support you need. You'll learn to translate strategic goals into clear decision inventories, define metrics that inform your choices, and hold your team accountable for results.

If you're an executive, this book provides the language to demand what you want: analytics that drives better, faster decisions. You'll learn what questions to ask, how to structure decision processes, and how to build decision quality as an organizational capability.

Your First Step

Before you begin Chapter 1, take five minutes for this exercise:

Write down one strategic goal your organization is pursuing. Something official: "Improve customer retention by 15%" or "Accelerate project delivery by 25%."

Now identify one important decision that should drive progress toward that goal. Not a vague management activity, but a specific choice someone needs to make repeatedly.

Finally, honestly assess: How is that decision made today? Is there a clear owner? A consistent process? Good information? High confidence?

The gap between your strategic goal and your decision reality is what this book helps you close.

You're about to learn how.

Chapter 1: You're Data Rich but Decision Poor

You've invested millions in analytics. Your organization has modern cloud platforms, sophisticated visualization tools, and talented data teams. Executives receive polished dashboards with traffic light indicators and trend lines. The data is real. The charts are clear. The metrics look impressive.

Yet when it comes time to make important decisions (which projects to fund, where to allocate staff, which customers to prioritize, how to balance operational risk and efficiency) those decisions rarely seem to flow from what's on the screen.

Instead, they happen through some opaque combination of institutional memory, political negotiation, personal relationships, and gut instinct. The dashboards exist. The metrics are tracked. But somehow they remain disconnected from the choices that actually shape organizational outcomes.

If you work in analytics, you've felt this frustration. You spend weeks building comprehensive analysis, only to watch executives glance at it for three minutes before making decisions based on other factors. You produce reports that are technically excellent but somehow never quite land. You answer the questions you're asked, but those questions don't seem to be the right ones.

If you're an operational leader, you've experienced the flip side. You know you should be "data driven," but when you look at the dashboards available to you, they don't clearly tell you what to do. You need to decide whether to hire for an open role or redistribute work, but the workforce metrics you have don't directly answer that question. You need to choose which improvement initiative to prioritize, but the project status reports leave the trade-offs murky.

If you're an executive, you've probably wondered why all this investment in analytics hasn't made decisions noticeably easier or better. You still find yourself making major choices based largely on experience and judgment, with data playing a supporting but not decisive role.

This isn't a failure of technology, talent, or effort. It's a structural problem. Organizations today have two things in abundance but lack the essential connection between them.

What Organizations Have

They have strategy documents. Mission statements, strategic plans, annual priorities, transformation roadmaps. These articulate where the organization wants to go: growth targets, market positions, operational excellence goals, quality aspirations, risk tolerances. The intentions are clear, often compelling.

They have data and analytics capabilities. Data warehouses, business intelligence platforms, reporting tools, dashboards, and increasingly sophisticated AI and machine learning. The ability to collect, store, process, and visualize information has never been greater. Organizations can measure almost anything they choose to measure.

What Organizations Lack

What's missing is the **clear, engineered bridge** between these two domains. The connection between strategic intent and the decisions people actually make. The link between the goals in planning documents and the daily or weekly choices that determine whether those goals are achieved or merely hoped for.

This gap manifests in predictable patterns you've probably witnessed:

Strategic goals that don't connect to Monday morning. Everyone can recite the priorities (improve customer retention, accelerate innovation, enhance operational efficiency) but it's unclear what anyone should do differently in response. The goals feel aspirational rather than actionable.

Decisions made repeatedly without clear criteria. The organization makes hiring decisions dozens of times per quarter, but each one is debated from scratch. Project funding happens through the same negotiation dance every budget cycle, with no systematic framework for evaluating trade-offs. Customer escalations are handled inconsistently because there's no shared logic for prioritization.

Metrics tracked but not acted upon. Dashboards show customer satisfaction declining, but no one has explicit accountability for improving it or clear authority to make changes. Project status turns red, but the response is more meetings, not different decisions. Operational metrics signal problems, but they don't trigger specific interventions.

Analytics that arrive too late or answer the wrong questions. By the time comprehensive analysis is ready, the decision has already been shaped by other conversations. Or the analysis is technically rigorous but addresses questions no one actually needs answered to make a choice. Or it presents findings without connecting them to specific decisions someone has authority to make.

Executives who know they should be data driven but aren't sure how. They dutifully review dashboards in meetings. They ask for more analysis when facing difficult choices. But data rarely settles debates or clarifies direction in the way they hoped it would. Often, the data just gives everyone more ammunition for positions they'd already formed.

If these patterns sound familiar, you're not alone. They exist in organizations across industries, regardless of how much has been invested in analytics infrastructure or how sophisticated the technical capabilities are.

Three Illusions That Keep Organizations Stuck

Organizations remain in this frustrating state because they operate under three persistent illusions about how data drives decisions.

Illusion One: Visibility Equals Impact

The first illusion is believing that making data visible automatically changes behavior.

Consider a large hospital system that built an elaborate emergency department dashboard showing wait times, patient volumes, and resource utilization across all facilities in real time. The dashboard was technically excellent. Leaders checked it regularly. Yet wait times continued to climb and patient satisfaction scores remained flat.

The problem wasn't the data or the visualization. The problem was that no one had defined the specific decisions this dashboard should support. Should facilities with high wait times receive additional staffing? Should patients be redirected to less busy locations? Should certain types of cases be triaged differently?

Without explicit decisions, the dashboard became a monitoring tool that made problems visible but didn't help anyone solve them. Everyone could see the numbers. No one knew what to do differently based on what they saw.

Visibility without decision structure creates awareness without agency. People see the numbers. They may even discuss them in meetings. But without a clear connection between what they're seeing and what they should do differently, the data becomes background noise.

Illusion Two: More Metrics Equal Better Management

The second illusion is that tracking more KPIs improves management effectiveness.

When leaders feel uncertain about performance, the instinct is to measure more things. If ten metrics aren't providing clarity, perhaps twenty will. If departments aren't aligned, perhaps they each need their own comprehensive scorecard.

A global manufacturing firm tracked over 150 different operational metrics across its supply chain. Each regional team had preferred indicators. Finance had its metrics, operations had different ones, and quality assurance tracked a third set entirely. When senior leaders asked for a clear picture of supply chain health, they received three different analyses that reached contradictory conclusions, all based on "the data."

The fundamental error was treating measurement as a substitute for decision clarity. Instead of first asking "What specific choices do we need to make about our supply chain?" the organization had allowed each function to independently decide what to measure, creating a measurement system disconnected from any cohesive decision framework.

More metrics without metric discipline creates proliferation without prioritization. Teams spend more time debating definitions and reconciling numbers than making decisions. And the metrics that actually matter (the ones that should inform specific choices) get lost in the noise.

Illusion Three: Color Coding Equals Strategy

The third illusion is that status indicators (red yellow green scorecards, up and down arrows, traffic lights) constitute strategic insight.

These visual shortcuts are appealing because they promise to distill complexity into simple signals. Green means good, red means bad, yellow means caution. The simplicity feels like clarity.

But color coding without decision context is just decoration. What action should follow from a red indicator? Does yellow mean "watch closely" or "intervene now"? Who is accountable for moving a metric from red to green, and by when?

A major retailer built a store performance dashboard that color coded dozens of metrics for each location. Store managers received weekly scorecards showing their performance across categories like sales, inventory, staffing, and customer service. The intent was to create transparency and accountability.

Instead, the system created confusion and gaming. Managers focused on whichever metrics happened to show red that week, neglecting others. Some discovered they could improve their color coding by adjusting how they scheduled staff or categorized inventory, without actually improving underlying performance. The dashboard's visual simplicity masked the absence of any coherent strategy about which trade-offs mattered and what managers should prioritize when metrics conflicted.

Why "More Data" Won't Fix This

These illusions persist because they let organizations feel like they're being data driven without doing the harder work of becoming decision driven.

Building dashboards is straightforward and visible. You can show executives the new BI platform, demonstrate the drill down capabilities, and present the comprehensive metric library. It looks like progress.

What's harder, and what most organizations skip, is the foundational work of identifying which decisions most directly influence strategic outcomes, assigning clear ownership for those decisions, defining what questions those decisions require answered, designing metrics that directly inform those specific questions, establishing processes that ensure metrics actually influence choices, and creating accountability for whether decisions produce expected results.

This foundational work is slower, messier, and often politically uncomfortable. It requires executives to be explicit about who decides what. It requires functions to align on shared definitions when they'd prefer their own. It requires analytics teams to push back on vague requests and insist on decision context. It requires everyone to acknowledge trade-offs rather than pretending every goal can be equally optimized.

But it's the only path to making data actually matter.

The Pattern That Keeps Repeating

Here's what the cycle looks like in most organizations:

Phase 1: Recognition that data isn't driving decisions

Leadership realizes that despite significant investment in analytics, decisions still happen largely through intuition and negotiation. There's a sense that "we should be more data driven."

Phase 2: Investment in better technology or more dashboards

The organization upgrades its BI platform, implements self service analytics, or launches a dashboard initiative. More people get access to more data with better visualizations.

Phase 3: Initial enthusiasm

The new capabilities look impressive. Executives explore dashboards in meetings. Analysts create sophisticated visualizations. Everyone talks about being data driven.

Phase 4: Gradual realization that nothing has fundamentally changed

A year later, decisions still happen the same way. The new dashboards are monitored but don't drive action. Analytics work still struggles to influence important choices. The gap between data availability and decision quality remains.

Phase 5: Repeat

Leadership concludes the problem must be inadequate technology, insufficient skills, or incomplete data. The cycle begins again with a new platform, new hires, or new data sources.

This cycle can repeat for years, consuming substantial budget and effort, without ever addressing the structural gap: the missing architecture that connects strategic goals to specific decisions, those decisions to the questions they require, those questions to decision critical metrics, and those metrics to reliable data and systematic action.

What's Actually Missing

The gap isn't about technology, data quality, or analytical sophistication. It's about decision architecture.

Most organizations have never explicitly designed their decision making as a capability. Decisions happen, certainly. Organizations make thousands of choices daily. But these decisions emerge from meetings, org charts, cultural norms, and individual initiative rather than from deliberate design.

No one maintains an inventory of the organization's most important decisions. No one has explicitly connected those decisions to the metrics that should inform them. No one has established clear ownership and consistent processes for high stakes, repeating choices. No one systematically reviews whether decisions are producing intended results and adjusts processes accordingly.

Decision making is treated as something that happens naturally when smart people have access to information, not as something that needs to be architected, governed, and continuously improved.

The result is that when you ask "How do we make decisions about X?" (whether X is capital allocation, workforce planning, customer priorities, or risk management) the honest answer is often: "It depends. Different people do it differently. The process isn't really documented. We kind of figure it out each time."

That's not a failure of any individual. It's the predictable result of never building the infrastructure that connects strategy to decisions to metrics to data.

A Preview of the Solution

The chapters that follow introduce the Strategy Layer: a structured approach to building that missing infrastructure.

The Strategy Layer is not another dashboard. It's not a new data platform. It's not an organizational restructuring. It's a framework and set of practices that make decision making a designed organizational capability.

At its core, the Strategy Layer creates an explicit chain. Strategic goals define what success looks like. Key decisions are the specific, repeatable choices that drive whether you achieve those goals. Essential questions are what you need to understand to make those decisions well. Decision critical metrics quantify the answers to those questions. Data and analytics feed those metrics with reliable information. Actions follow systematically from decisions. Outcomes close the loop by revealing whether it's working.

Each link in this chain is visible, owned, and improvable. When something isn't working (decisions aren't improving outcomes, metrics aren't informing decisions, or analytics isn't answering the right questions) the architecture makes the problem visible rather than hiding it in organizational complexity.

Building the Strategy Layer requires discipline. You can't track every possible metric or support every possible decision. You must make choices about what matters most. You must assign accountability where it's currently diffuse. You must insist on connecting work to decisions rather than letting requests flow based on who asks loudest.

But organizations that build this foundation find that their analytics investments finally pay off. Data begins to drive decisions not because the data got better, but because the connection between data and decisions was deliberately engineered.

What This Means for You

If you're an analyst or analytics manager, this book provides the framework you need to shift from being a report factory to being a strategic decision partner. You'll learn how to insist on decision context before starting work, how to prioritize based on decision impact rather than stakeholder volume, and how to structure analysis that executives actually use to make choices.

If you're an operational leader, this book shows you how to get the decision support you've been missing. You'll learn how to translate your strategic goals into clear decision inventories, how to define metrics that actually inform your choices, and how to hold your analytics team accountable for supporting decisions rather than just producing information.

If you're an executive, this book gives you the language and frameworks to demand what you've been wanting: analytics that actually drives better, faster decisions. You'll learn what questions to ask when someone presents analysis, how to structure decision processes so data plays a decisive role, and how to build decision quality as an organizational capability that creates sustained competitive advantage.

The path forward isn't more dashboards. It isn't better visualization tools. It isn't additional data sources or more sophisticated AI models, though all of those can be valuable when applied within the right structure.

The path forward is building the Strategy Layer that gives your data, your analytics, and your decision making purpose and direction.

The next chapter introduces that framework in detail. But first, you need to see clearly why your current approach (no matter how much you've invested) can't close the gap without it.

You've already recognized the problem. You're living it. Now it's time to understand the solution.

Mini Exercise: Recognize the Pattern in Your Organization

Before moving forward, take ten minutes to identify where these patterns show up in your work:

1. Name one strategic goal your organization is pursuing.

Write it down as it's officially stated: "Improve customer retention," "Accelerate digital transformation," "Enhance operational efficiency," or whatever your leadership emphasizes.

2. Identify one important decision that should drive progress toward that goal.

Not a vague management activity, but a specific choice: "Which at risk customers to target for retention programs," "Which technology projects to fund next quarter," "Where to allocate operational improvement resources."

3. Honestly assess: How is that decision made today?

Is there a clear owner who has authority to decide? Is there a consistent process, or does it vary each time? What information informs the decision: metrics, analysis, or mostly discussion and judgment? How confident are decision makers when they choose? Do you know whether past decisions produced the intended results?

4. Note the gap.

The distance between what you wrote in step 1 (strategic goal) and what you described in step 3 (decision reality) is the gap this book helps you close.

Keep these notes. As you learn to build the Strategy Layer in subsequent chapters, you'll return to this decision and transform how it's made.

The gap you just documented isn't unique to your organization. It's nearly universal. The difference is that you're now going to do something about it.

Chapter 2: The Strategy Layer Framework

You've seen the problem: organizations drowning in data but struggling to make decisions. The gap between strategic intent and daily reality persists despite massive investment in analytics. The three illusions (visibility equals impact, more metrics equal better management, color coding equals strategy) keep organizations stuck in cycles that consume resources without improving outcomes.

Now it's time to understand the solution.

This chapter introduces the Strategy Layer: a structured framework that bridges the gap between strategy and execution by making decisions (not data) the organizing principle of your analytics work.

What the Strategy Layer Is

The Strategy Layer is the explicit architecture that connects strategic goals to the specific decisions that drive those goals, and connects those decisions to the metrics, data, and actions that make them work.

Think of it as the operating system for strategic execution. Your strategy documents provide direction (where you want to go). Your data platforms provide capability (what you can measure and analyze). The Strategy Layer is the connective tissue that ensures direction and capability actually work together to produce better decisions.

Without the Strategy Layer, you have strategic plans on one side and operational data on the other, with only informal connections, personal relationships, and institutional memory bridging the gap. Some decisions work out, others don't, and the organization struggles to learn systematically from either.

With the Strategy Layer, you have an explicit chain of logic. Strategic goals identify what matters. Critical decisions determine whether you achieve those goals. Essential questions define what you need to know to decide well. Decision critical metrics quantify the answers. Data and analytics provide reliable information. Actions follow systematically from decisions. Outcomes reveal whether it's working.

Each link is visible, owned, and improvable.

The Strategy Layer Flow

Here's how the Strategy Layer connects its components:

1. **Strategic Goals** → Define what success looks like
2. **Critical Decisions** → Identify specific, repeatable choices driving goals
3. **Essential Questions** → Determine what you need to know to decide
4. **Decision-Critical Metrics** → Measures that inform and trigger action
5. **Data & Analytics** → Deploy information sources and analytics
6. **Actions** → Execute what changes based on decisions
7. **Outcomes** → Measure results that validate or challenge your approach
8. **Feedback Loop** → Refine goals and decisions based on outcomes

This flow forces clarity at every level. You can't define decision critical metrics without knowing what questions matter. You can't identify critical questions without knowing what decisions you're trying to make. You can't prioritize decisions without clear strategic goals.

The framework also reveals dependencies. If your metrics aren't improving outcomes, either the metrics are measuring the wrong things, the decisions aren't being informed by metrics, the actions aren't following from decisions, or the strategy itself may be flawed. The Strategy Layer makes these failure modes visible.

How This Differs From What You Have Now

Most organizations already have pieces of what the Strategy Layer provides. They have strategic planning processes. They have decision making practices. They have analytics capabilities. What they lack is the explicit integration that makes these pieces work as a system.

Your strategic plan identifies goals but typically doesn't specify the critical decisions that determine whether you'll achieve them. It says "improve customer retention by 15%" but doesn't inventory the repeating choices (which customers to target, what retention tactics to deploy, how to allocate retention investment) that actually drive retention outcomes.

Your organization makes decisions constantly, but most aren't cataloged, documented, or deliberately designed. There's no central view of which decisions matter most, who owns them, what information they need, or whether they're being made well. Important choices happen through informal processes that vary by person and circumstance.

Your metrics exist in dashboards and reports, but most aren't explicitly tied to specific decisions. No one can easily answer: "Which metric informs which decision, and who's accountable for acting when that metric changes?" Metrics proliferate based on what seems interesting or what someone once requested, not based on decision requirements.

Your analytics team works hard but primarily in reactive mode, building what's requested without clear frameworks for prioritization. When someone asks for analysis, the team often can't answer: "Which strategic decision does this support?" because that connection was never made explicit.

The Strategy Layer integrates these disconnected pieces by starting with decisions as the organizing principle and working systematically outward from there.

Why Decisions Are the Unit of Design

Data doesn't act. Metrics don't act. Dashboards don't act. People act, and they act through decisions.

You don't "act on data" in some general sense. You decide whether to fund a project or kill it. You decide whether to hire for an open role or redistribute responsibilities. You decide whether to enter a new market or double down on existing ones. You decide whether to accept a risk or invest in mitigation. Every action that matters flows through a decision.

This makes decisions the natural unit of design for connecting strategy to execution.

Decisions provide natural prioritization. Not all decisions are equally consequential. Some happen frequently and compound over time. Others are rare but high stakes. Some directly influence strategic outcomes; others are necessary but tangential. By making decisions visible, you can prioritize where to invest analytics resources based on decision impact rather than whoever asks loudest.

Decisions clarify ownership and accountability. Every important decision should have a clear owner (someone who ultimately makes the choice and is accountable for whether it works). When decisions are invisible, ownership is ambiguous. When decisions are cataloged, you can assign explicit accountability and ensure each gets the support it needs.

Decisions connect directly to strategy. Strategic goals become real through repeating, high impact decisions. If your strategy is to grow market share, which decisions actually drive that? Pricing decisions, customer acquisition decisions, retention decisions, product investment decisions. Making these explicit lets you see whether your decision infrastructure supports your strategy or undermines it.

Decisions expose gaps. When you inventory decisions, you often discover that critical choices aren't being made systematically at all. They happen inconsistently across the organization, through informal processes that don't scale, or they're deferred until crisis forces rushed choices. Exposing these gaps is the first step toward addressing them.

The Core Components

Building the Strategy Layer requires developing several interconnected components. Think of these as the architecture that makes systematic decision support possible.

1. Decision Inventory

The decision inventory is your catalog of important decisions, capturing the characteristics that let you understand, prioritize, and improve them.

For each decision, you document:

- What choice is being made (framed as selection between clear options)
- Who owns it (a role or person with authority, not a committee)
- How often it occurs (daily, weekly, monthly, quarterly, annually, or event triggered)
- Who provides input (stakeholders and functions that should be consulted)
- What strategic goal it serves (the connection to your strategy)
- How it's made today (current process, pain points, information used)

This inventory transforms invisible, informal decision making into visible, manageable components of your operating architecture. It becomes the backbone for everything else in the Strategy Layer.

2. Metric Map

The metric map shows which metrics inform which decisions, creating explicit connections between measurement and action.

For each decision critical metric, you specify:

- Which decision it informs (from your inventory)
- What question it answers (the specific information need)
- How it's defined and calculated (precise specification)
- What data sources feed it (where information comes from)
- Who owns it (accountable for accuracy and maintenance)
- What thresholds trigger action (ranges that map to different choices)

This map prevents metrics from proliferating without purpose and ensures that the metrics you maintain actually serve decisions rather than just monitoring performance.

3. Governance Structure

Governance defines how the Strategy Layer itself is maintained and evolved as strategy and circumstances change.

Key governance elements include:

- Decision review forums where owners examine whether critical decisions are being made well
- Metric change control so decision critical metrics are updated deliberately, not casually
- Request intake requiring decision context so analytics work connects to documented decisions
- Strategy Layer stewardship with explicit accountability for maintaining the architecture
- Periodic health assessments revealing whether the Strategy Layer is improving decision quality

Governance transforms the Strategy Layer from a one time initiative into a sustainable operating capability.

4. Decision Support Artifacts

These are the tools and templates that operationalize the Strategy Layer in daily work:

- Decision briefs that structure how options and evidence are presented to decision makers
- Intake forms that require decision context before analytics work begins
- Review templates for examining whether decisions produced expected outcomes
- Metric definition standards ensuring consistency across the organization

These artifacts make the Strategy Layer tangible (not abstract concepts but practical tools people use when making and supporting decisions).

What the Strategy Layer Is Not

Understanding what the Strategy Layer isn't helps clarify its purpose and prevent misapplication:

Not just a dashboard. Dashboards might be part of the Strategy Layer, but they're artifacts, not the architecture itself. A dashboard without clear connection to specific decisions is just information display.

Not a data platform. You need data infrastructure to operationalize the Strategy Layer, but the Strategy Layer sits conceptually above your platforms, defining what they should enable rather than being defined by their capabilities.

Not a new org structure. You don't need to reorganize to implement the Strategy Layer, though you do need clarity about who owns which decisions. It works within existing structures by making decision architecture explicit.

Not a one time project. It's an operating model that requires ongoing governance, maintenance, and adaptation. Think of it like financial controls or quality management (a continuous capability, not a program with an end date).

Not just for analytics teams. While analytics teams play crucial roles, the Strategy Layer is fundamentally about how leaders make decisions. It requires partnership between decision owners, analytics professionals, and functional experts who execute.

Not a replacement for judgment. The Strategy Layer helps decisions get informed by better information, but it doesn't automate judgment out of the process. Leaders still need to weigh trade-offs, consider factors beyond data, and make calls under uncertainty.

How the Strategy Layer Solves the Three Illusions

Remember the three illusions from Chapter 1 that keep organizations stuck? The Strategy Layer directly addresses each:

Solving Visibility ≠ Impact

The Strategy Layer transforms visibility into impact by connecting every metric to a specific decision with clear ownership and defined thresholds.

That hospital ED dashboard showing wait times? In a Strategy Layer approach, it's explicitly designed to support specific decisions: How to allocate nursing staff across shifts. When to activate surge protocols. Whether to adjust triage processes. Each decision has an owner who reviews relevant metrics on a cadence that matches decision frequency and knows what actions to take based on what the data shows.

Visibility becomes impact when it's coupled with decision authority and clear action paths.

Solving More Metrics ≠ Better Management

The Strategy Layer solves metric proliferation by making every metric justify its existence through connection to a decision.

When someone proposes a new metric, you ask: "Which decision in our inventory does this inform?" If they can't articulate that connection, the metric doesn't get added (or gets classified as a lower priority health or vanity metric that doesn't warrant the same maintenance investment as decision critical measurements).

This discipline creates focused measurement systems where fewer metrics get better maintenance and actually influence choices, rather than sprawling collections where most metrics are calculated but rarely used.

Solving Color Coding ≠ Strategy

The Strategy Layer replaces decorative indicators with decision frameworks that specify what different metric values mean and what actions they should trigger.

That retail store scorecard with color coded performance? In a Strategy Layer approach, each metric is tied to specific store operations decisions (staffing adjustments, inventory reordering, customer service focus areas). Red doesn't just mean "bad"; it means "trigger this specific intervention." Yellow doesn't just mean "caution"; it means "review at weekly operations meeting and decide whether to adjust."

Color coding becomes strategy when it's embedded in decision processes with clear ownership and explicit action rules.

A Simple Example: Workforce Planning

To make this concrete, here's how the Strategy Layer transforms one common organizational challenge: workforce planning.

Without the Strategy Layer:

HR tracks headcount, turnover rates, time to fill for open positions, and cost per hire. Leaders review these numbers in monthly HR reports. When turnover seems high or hiring is slow, they ask for more analysis. Decisions about where to hire, what to pay, and when to promote happen through separate processes that rarely reference the metrics explicitly. Each business unit approaches workforce planning differently.

With the Strategy Layer:

Strategic Goal: Build and retain the capabilities needed for our growth strategy

Critical Decisions (documented in inventory):

1. Where to allocate hiring budget across roles and locations (quarterly)
2. Which open positions to fill versus leaving vacant (monthly)
3. What compensation to offer for hard to fill roles (case by case)
4. Which high performers to invest in retaining (continuous)

Essential Questions:

- Which roles have highest attrition and highest replacement cost?
- Where is turnover among high performers versus low performers?
- What market compensation rates apply to critical skills?
- Which roles are blocking strategic initiatives due to vacancies?
- What retention tactics have worked when we've tried them before?

Decision Critical Metrics:

- Attrition rate by role and performance level
- Time to fill by role criticality
- Compensation competitiveness index by role
- Strategic project delays attributed to staffing gaps
- Retention program ROI by intervention type

Data Sources:

- HRIS for headcount and movement
- Performance management system for ratings
- Compensation benchmarking services
- Project management system for initiative staffing
- Exit interview data for attrition drivers

Actions:

- Quarterly hiring budget allocated to roles with highest strategic impact and recruiting difficulty
- Monthly review triggers either hiring approval or work redistribution
- Compensation exceptions approved based on market data and role criticality
- Targeted retention programs deployed for high value at risk employees

Outcomes:

- Reduced time to fill for critical roles (from 90 to 60 days)
- Lower regrettable attrition (high performers leaving)
- Better strategic project staffing
- More efficient use of hiring budget

Governance:

- Monthly workforce planning forum reviews decisions and metrics
- Quarterly metric definitions updated based on what's working
- Annual strategic alignment check ensures decisions still serve goals

This example shows how the Strategy Layer integrates previously disconnected pieces (strategic goals, operational decisions, measurement systems, analytics work, and execution) into a coherent system where each component has clear purpose and accountability.

Starting Points and Scope

You don't build the Strategy Layer for your entire organization at once. That's overwhelming and produces more documentation than insight.

Start with one strategic goal. Pick something that genuinely matters (if you don't make progress here, people notice). "Improve customer retention," "accelerate project delivery," "enhance operational efficiency," or whatever your leadership emphasizes as critical.

Identify three to five decisions that drive that goal. Not every decision related to the goal, just the repeating, high impact choices where better information would consistently produce better outcomes.

Build the full Strategy Layer for those decisions. Document them in your inventory, define their decision critical metrics, map data sources, establish governance, and actually use this framework to make those decisions over several cycles.

Prove it works. Measure whether decisions get made faster, with higher confidence, and produce better results. Document specific examples of improved outcomes.

Then expand. Once you've proven value in one domain, extend to additional strategic goals using patterns you've learned. Build enterprise infrastructure as multiple domains mature.

This focused approach lets you demonstrate results quickly while learning what works in your organizational context. It's faster than enterprise wide rollout attempts that often stall in endless planning.

What Changes When You Have This

Organizations with functioning Strategy Layers operate differently from organizations without them:

Meetings have explicit purpose. Every significant meeting starts with "We're here to decide X" rather than "Let's talk about Y." When no decision needs to be made, that's acknowledged and the meeting is canceled or shortened.

Analytics work is prioritized rationally. When an analyst receives more requests than capacity allows, decisions from the inventory take priority. Requests that can't connect to documented decisions get deferred or redirected to self service tools.

Executives get decision support, not just data. Instead of presentations with dozens of charts and ambiguous implications, executives receive decision briefs: here's the choice, here are the options, here's what the evidence shows, here's what we recommend.

Metrics have clear owners and action thresholds. When a decision critical metric moves significantly, someone knows they're accountable and what they should do. There's no ambiguity about whether action is needed or who should take it.

Trade-offs are explicit, not hidden. When metrics point in different directions (cost optimization versus quality improvement, growth versus profitability, speed versus risk reduction) the organization has frameworks for making those trade-offs explicit rather than letting them get resolved through organizational politics.

Learning compounds over time. Because decisions are documented and outcomes are tracked, the organization builds knowledge about what works. Today's decisions benefit from lessons learned in previous iterations.

New leaders can understand decision landscape quickly. When someone joins the organization or moves to a new role, the Strategy Layer provides clear documentation of critical decisions in their domain, how they're currently made, and what metrics inform them.

These changes don't happen overnight. They accumulate as the Strategy Layer matures. But organizations that build this infrastructure create measurable advantages over competitors still operating without it.

The Journey Ahead

The remaining chapters provide the detailed frameworks and practical guidance for building your Strategy Layer:

Chapters 3 to 6 teach you to build the core components (decision inventories, decision critical metrics, decision briefs, and transformed analytics intake processes). Each chapter includes step by step implementation guides, filled examples, and troubleshooting advice.

Chapters 7 to 8 show you how to apply the Strategy Layer to common organizational challenges (cross-functional alignment and AI deployment) where the framework creates clarity that's otherwise elusive.

Chapter 9 provides the roadmap for scaling from initial success to enterprise capability, including governance patterns, maturity assessment, and leadership development.

Throughout, you'll find tools you can use immediately: templates, checklists, worksheets, and examples drawn from realistic organizational scenarios. This isn't primarily a book to read (it's a book to use).

The Strategy Layer isn't theoretical. It's practical infrastructure you can start building Monday morning.

What Success Looks Like

You'll know the Strategy Layer is working when:

Decisions that used to take weeks now take days because the information needed is readily available and decision processes are clear rather than ad hoc.

Executives consistently reference the Strategy Layer in their conversations ("Which decision in our inventory does this impact?" becomes natural language, not forced framework adoption).

Analytics teams feel their work matters because they see it directly influence important choices rather than disappearing into unused reports.

Strategic initiatives execute more reliably because the decisions they depend on actually happen, informed by appropriate evidence rather than political negotiation.

Cross-functional conflicts decrease because shared decision frameworks and explicit trade-off rules replace territorial battles over whose metrics matter more.

New capabilities get adopted faster because AI and advanced analytics are deployed within clear decision contexts rather than as solutions searching for problems.

Your organization learns systematically from outcomes, improving decision quality over time rather than repeating the same patterns regardless of results.

This is what it means to close the gap between strategy and execution. Not through better strategy documents or more sophisticated analytics, but through the deliberate architecture that connects them.

You're ready to start building.

Reflection Exercise: Where to Begin

Before diving into detailed implementation in the next chapters, identify your starting point:

1. Choose one strategic goal from your organization's current priorities. Pick something concrete: "Increase customer retention by 15%," "Reduce project delivery time by 25%," "Improve employee engagement to top quartile."

2. Brainstorm five to ten decisions that influence whether you'll achieve that goal. Think about choices that get made repeatedly (weekly, monthly, or quarterly) not one time strategic moves.

3. Narrow to your top 3 based on these criteria:

- Which decisions happen frequently enough to benefit from systematic support?
- Which have highest impact on the strategic goal?
- Which currently suffer from inadequate information or unclear process?

4. For each of those 3 decisions, note:

- Who makes this decision today? (Ideally one person, not a committee)
- How often does it occur?
- What makes it difficult or contentious?
- What information would make it easier to decide well?

Keep these notes. The next chapter teaches you to build a formal decision inventory, and these three decisions will be your starting point. You'll document them properly, define their metrics, and establish processes to support them.

You've just identified where your Strategy Layer begins.

The gap between strategy and execution is about to close.

Chapter 3: Building Your Decision Inventory

Most organizations manage portfolios. They catalog projects, track investments, maintain inventories of systems, vendors, and risks. Yet few organizations maintain an explicit inventory of their most consequential decisions.

This gap is puzzling. If decisions are how strategy becomes real, if they're the primary mechanism through which organizations allocate resources and create value, why aren't they cataloged and deliberately improved the same way projects and assets are?

This chapter makes decisions the primary organizing unit of your Strategy Layer. You'll learn to build a decision inventory that transforms choices from invisible, informal processes into visible, manageable components of your organizational architecture.

Why Start With Decisions

Before you can build metrics, design dashboards, or prioritize analytics work, you need clarity about what decisions you're trying to support. Without this foundation, you're building in the dark.

Decisions provide natural prioritization. Not all choices are equally consequential. Some happen frequently and compound over time. Others are rare but high stakes. Some directly influence strategic outcomes; others are necessary but peripheral. By inventorying decisions explicitly, you can prioritize where to invest resources based on decision impact rather than who asks loudest.

Decisions clarify ownership and accountability. Every important choice should have a clear owner. Someone must make the call and be accountable for whether it works. When decisions are invisible, ownership is ambiguous. When they're inventoried, you can assign explicit accountability and ensure each gets the support it needs.

Decisions connect directly to strategy. Strategic goals become real through repeatable, high impact choices. If your strategy is to grow market share, which decisions actually drive that? Customer acquisition, pricing, product investment, market entry. Making these explicit reveals whether your decision infrastructure supports your strategy or undermines it.

Decisions expose gaps. When you catalog choices systematically, you often discover that critical decisions aren't being made at all. They happen inconsistently, through informal processes that don't scale, or they're deferred until crisis forces rushed action. Exposing these gaps is the first step to addressing them.

What Belongs in Your Inventory

A decision inventory is a structured catalog of important decisions, capturing characteristics that let you understand, prioritize, and improve them. Think of it as the master list of choices that shape whether your organization succeeds.

Your inventory should focus on decisions that meet three criteria:

Strategic importance. The decision directly influences strategic goals. Poor choices here create meaningful setbacks; good choices create measurable progress. These aren't routine operational tasks but consequential choices where outcomes matter.

Repeatability. The decision happens regularly enough to benefit from systematic support. Weekly, monthly, or quarterly decisions compound their impact over time. One time strategic moves are important but don't need the ongoing infrastructure the Strategy Layer provides.

Decision quality gap. There's visible room for improvement. The choice currently suffers from inadequate information, unclear process, inconsistent execution, or low confidence. If decisions are already being made well with high certainty, they may not need Strategy Layer support.

Not every decision meets all three criteria. That's fine. Your inventory focuses on the high value choices where systematic support will create the biggest difference.

Essential Elements to Capture

For each decision in your inventory, document these core elements:

Decision Name and Description

State clearly what choice is being made. Frame it as selection between options, not as general management activity.

Weak: "Manage customer relationships"**Strong:** "Decide which at-risk enterprise customers to target for retention programs each quarter"

Weak: "Optimize staffing"**Strong:** "Decide how to allocate nursing staff across units and shifts based on daily census and acuity"

Weak: "Portfolio management"**Strong:** "Decide which IT projects to continue funding versus defer or cancel at quarterly portfolio reviews"

The description should be specific enough that someone unfamiliar with your organization could understand what's at stake.

Decision Owner

Name the role or person who ultimately makes this choice. Not a committee, not a department. One individual with authority and accountability.

If the decision requires input or consensus from others, note that separately. But identify who has final authority. If you can't name an owner, that's a red flag suggesting the decision happens inconsistently or not at all.

Good ownership: "VP of Operations"**Acceptable:** "Regional Operations Manager for Northeast"**Problem:** "Leadership team" (too diffuse)**Problem:** "Sales and Marketing jointly" (split accountability)

When ownership is genuinely shared, specify who breaks ties or makes the final call when consensus fails.

Frequency and Timing

How often does this decision occur? Be specific about timing, not just frequency.

Examples:

- "Weekly on Monday mornings during operations review"
- "Quarterly in advance of budget cycles, typically mid-month of first quarter month"
- "Monthly within first week as previous month data becomes available"
- "Triggered when project variance exceeds 15% threshold, typically 3-5 times per quarter"

Understanding frequency matters for several reasons. High frequency decisions benefit from more systematic processes and potentially automation. Low frequency decisions need more deliberate structure since there's less opportunity to learn through repetition. Some decisions feel ad hoc but actually recur in patterns that can be supported systematically.

Stakeholders and Input Sources

Who needs to provide information or perspective? Which functions or roles should be consulted? This isn't about consensus decision making. The owner still owns the choice. But it's about ensuring the decision gets informed by relevant expertise and considers impacts across the organization.

Example for capital allocation decision:

- Finance provides budget analysis and ROI projections
- Business unit leaders provide strategic rationale and resource requests
- IT assesses technical dependencies and infrastructure requirements
- Risk evaluates control adequacy and compliance implications
- Strategy validates alignment with enterprise priorities

List stakeholders who genuinely influence the decision through their input, not everyone who might be interested.

Current State Assessment

How is this decision made today? Document the reality, not the ideal.

Questions to answer:

- Is there a formal process, or does it happen informally?
- What information is currently considered?
- How long does it typically take from recognizing the need to decide until action happens?
- What makes this decision difficult or contentious?
- Where does the process break down or cause frustration?

Be honest about pain points. The goal isn't to justify current approaches but to identify where improvement would create value.

Example: "Project funding decisions currently happen through quarterly portfolio reviews. Business sponsors present cases to leadership committee. Discussions last 2-3 hours per session and often end without clear resolution. Projects with executive champions tend to get funded regardless of business case strength.

No consistent criteria for evaluating trade-offs between strategic value, cost, risk, and resource availability. Decisions frequently revisited as circumstances change, creating uncertainty for project teams."

This assessment establishes your baseline and helps prioritize which decisions need Strategy Layer support most urgently.

Strategic Connection

Which strategic goal or goals does this decision influence? This connection anchors the decision in your Strategy Layer and helps with prioritization.

Be specific about the link between the decision and the outcome:

"Hiring allocation decisions directly influence our talent acquisition strategy (Goal: Build technical capabilities for platform modernization) by determining whether we staff critical roles fast enough to maintain project timelines."

"Customer retention investment decisions drive our growth strategy (Goal: Increase recurring revenue 20% annually) by reducing churn and expanding wallet share among existing accounts."

Decisions that connect to multiple strategic goals warrant extra attention. Decisions that don't clearly connect to any strategic priority may not belong in your inventory.

Building Your First Inventory

Don't try to catalog every decision across your entire organization. Start focused.

Step 1: Choose Your Domain

Pick one strategic goal or business area where you have genuine authority and where improved decisions would create visible value.

Good starting domains:

- IT portfolio decisions for organizations with substantial technology investment

- Workforce planning decisions for labor intensive operations
- Customer investment decisions for subscription or relationship based businesses
- Operational capacity decisions for service delivery organizations
- Risk acceptance decisions for regulated industries

The key is choosing something that matters and that you can actually influence. An analyst or analytics manager might start with decisions in their supported business unit. An operational leader might start with decisions in their domain. An executive might start with enterprise level choices they directly own.

Step 2: Brainstorm Decision Candidates

With your domain chosen, generate potential decisions to include. Use these prompts:

"What choices do we make repeatedly that shape outcomes?" Think about weekly, monthly, or quarterly choices, not one time strategic moves.

"Where do we struggle with inconsistent decision making?" Identify areas where similar situations get handled differently depending on who's involved or what day it is.

"What decisions currently rely heavily on gut instinct that could benefit from better information?" Look for high stakes choices made with low confidence or frequent second guessing.

"Where do we see poor decision outcomes that we'd like to improve?" Consider decisions that often lead to rework, waste, missed opportunities, or unintended consequences.

Get these on a whiteboard or shared document. Aim for 15 to 30 candidates initially. Don't filter yet. Just capture the possibilities.

Step 3: Prioritize Ruthlessly

You can't build Strategy Layer support for 30 decisions simultaneously. Narrow to your top 3 to 5 using these filters:

Impact on strategic goals. Which decisions most directly influence whether you achieve your strategic objectives? Rate each candidate on a simple scale: high, medium, low strategic impact.

Decision frequency. How often does each occur? Higher frequency often means higher compound value from improvement, though some low frequency decisions are so consequential they warrant inclusion anyway.

Current quality gap. Where is the distance between current and potential decision quality largest? Rate the pain points: severe, moderate, minor.

Feasibility. Can you actually access the data and stakeholders needed to support this decision? Some important decisions may depend on information you can't currently get or involve politics beyond your influence. Start with decisions you can reasonably improve.

Score or discuss these factors, then select your top 3 to 5. These become your initial inventory.

Step 4: Document Each Decision

For your prioritized decisions, complete the full documentation using the elements described earlier:

- Decision name and description
- Owner
- Frequency and timing
- Stakeholders and inputs
- Current state assessment

- Strategic connection

Use a simple format. A spreadsheet works fine. A table in a shared document works fine. Fancy tools can come later. What matters is having clear, accessible documentation everyone can reference.

Step 5: Validate With Decision Owners

Before considering your inventory complete, review it with the people who actually make these decisions. Do they recognize the descriptions? Do they agree with the ownership assignments? Do the pain points resonate?

This validation serves two purposes. It ensures accuracy and it begins building buy in. When decision owners help shape their own decision documentation, they're more likely to engage with the Strategy Layer support you'll build around those decisions.

Common Patterns Across Decision Types

While every decision is unique to its context, certain patterns recur across organizations. Recognizing these patterns helps you structure your inventory and apply lessons learned across similar decision types.

Investment and Portfolio Decisions

These govern how organizations allocate capital, budget, and major resources. Project portfolio funding, product line investment, market entry, technology platform selection, and major capital expenditures all fall here.

Common characteristics:

- Periodic cadence (quarterly or annually)
- High stakes (significant resources at risk)
- Multiple competing options (portfolio optimization)

- Long term consequences (decisions difficult to reverse quickly)
- Cross functional impact (affects multiple parts of organization)

Typical pain points:

- Insufficient visibility into resource constraints
- Inconsistent evaluation criteria across options
- Political factors overriding business logic
- Lack of follow through on portfolio decisions (projects approved but not adequately resourced)

Operational Capacity Decisions

These determine how work gets done day to day or week to week. Staffing allocation, production scheduling, inventory positioning, capacity routing, and service prioritization fit this category.

Common characteristics:

- High frequency (daily or weekly)
- Shorter time horizons (decisions affect near term operations)
- Resource constrained (finite capacity to allocate)
- Cumulative impact (individual decisions modest, pattern significant)
- Information intensive (require current data on demand and capacity)

Typical pain points:

- Data latency (information arrives too late to inform decisions)
- Forecasting difficulty (demand or capacity uncertainty)

- Local optimization (individual decisions rational but create system inefficiency)
- Inadequate feedback loops (outcomes not tracked to improve process)

Risk and Control Decisions

These manage how organizations identify, evaluate, and respond to threats. Escalation decisions, exception approvals, control investment choices, and incident response all fall here.

Common characteristics:

- Event triggered (happen when situations arise, not on schedule)
- Asymmetric consequences (errors more costly than successes)
- Uncertainty heavy (probabilistic rather than deterministic)
- Trade off intensive (balance protection against agility)
- Compliance constrained (regulatory requirements limit options)

Typical pain points:

- False positive fatigue (too many alerts dilute attention)
- Inconsistent response (similar situations handled differently)
- Unclear escalation thresholds (when to elevate versus handle locally)
- Poor cost benefit frameworks (over invest in low risk or under invest in high risk)

Customer and Market Decisions

These shape how organizations interact with customers and markets. Account prioritization, pricing strategies, customer retention investment, market segment targeting, and service level differentiation fit here.

Common characteristics:

- Revenue impacting (directly affect top line)
- Competitive context (influenced by market dynamics)
- Segmentation opportunities (different approaches for different customer types)
- Lifetime value considerations (short term and long term trade offs)
- Data availability challenges (customer behavior information often incomplete)

Typical pain points:

- Segment definition ambiguity (unclear how to group customers)
- Limited predictive capability (don't know which customers will churn or expand)
- Sales and operations misalignment (different functions optimize differently)
- Inadequate win loss analysis (don't learn systematically from outcomes)

Understanding which pattern your decisions follow helps you anticipate challenges and borrow solutions from similar contexts.

Red Flags and How to Address Them

As you build your inventory, watch for these warning signs:

"We don't really have decision owners"

If you can't identify clear owners, decisions probably aren't being made systematically. This is the problem you're trying to solve, not a reason to skip the inventory.

Address it: Propose ownership based on who should logically own the decision given their role and accountability. Document this as "proposed owner" and use the Strategy Layer work as opportunity to clarify and formalize what was previously ambiguous.

"These decisions happen in committees"

Committee involvement is fine for input and consensus building. But someone must have authority to make the final call when consensus fails. If genuine shared authority exists, specify how disagreements get resolved.

Address it: Identify who chairs the committee or has tiebreaker authority. That's your owner. Others become stakeholders who provide input.

"We make thousands of these decisions; we can't inventory them all"

You're not trying to inventory every decision. Only consequential, repeatable choices where systematic support would improve outcomes.

Address it: Use the prioritization criteria (strategic impact, frequency, quality gap) to narrow to the vital few rather than attempting comprehensive coverage.

"This decision happens differently in every region/business unit"

Variation isn't inherently bad. But if the same type of decision should be made consistently and isn't, that's a coordination problem the Strategy Layer can help solve.

Address it: Document how the decision should be made, acknowledging that local execution may vary. Or document it separately for each context if the decisions are genuinely different due to legitimate local factors.

"Nobody will use this inventory"

Inventories that sit on shelves collecting dust are useless. The inventory has value only when it's actively used to prioritize analytics work, structure decision processes, and track improvement.

Address it: Build the inventory to solve immediate problems. Use it to push back on vague analytics requests ("Which decision in our inventory does this support?"). Reference it in decision meetings ("Let's review how we documented this decision process"). Make it a working tool, not documentation theater.

Using Your Inventory Strategically

Once built, your decision inventory becomes the backbone of your Strategy Layer. Here's how to leverage it:

Prioritizing Analytics Work

When your analytics team receives more requests than capacity allows, the inventory provides rational prioritization. Requests that support high impact decisions from the catalog take priority. Requests that don't connect to inventoried decisions either get deferred or prompt a conversation about whether this is actually a critical decision that should be added.

A technology company used this approach to transform analytics capacity allocation. Previously, they responded to requests first come, first served, with the most persistent requesters getting the most attention. After building a decision inventory focused on strategic goals for operational efficiency and margin improvement, they redirected analytics work toward decisions about production scheduling, material sourcing, and product mix. Work that didn't support catalog decisions was politely declined or pushed to self service tools. Within six months, analytics became measurably more influential in decisions that actually moved the business.

Identifying Blind Spots

Looking across your inventory often reveals decisions that matter strategically but lack adequate information support. These blind spots become candidates for focused improvement.

Perhaps you have a critical decision about which customer segments to target for expansion, but you don't have reliable data on segment profitability or growth potential. Or you have important decisions about technical talent retention, but you lack metrics that predict attrition risk. Making these gaps visible creates a clear agenda for data strategy investments.

Aligning Cross Functional Work

Many strategic outcomes depend on decisions made by different functions that need coordination. The inventory makes these dependencies visible.

If your customer experience strategy depends on both sales decisions (which prospects to pursue) and service decisions (which customer issues to prioritize), seeing both in the inventory prompts conversations about alignment. Do these decisions use compatible metrics and assumptions? Are trade offs between acquisition and retention explicit? The inventory becomes a coordination mechanism.

Tracking Decision Maturity

As you improve how decisions are made through better metrics, clearer processes, and stronger governance, you can track maturity in the inventory. This creates a roadmap for building decision capability across the organization.

You might rate each decision on dimensions like clarity of ownership (1 to 5), availability of decision critical metrics (1 to 5), and stakeholder confidence in the process (1 to 5). Over time, you see which decisions are improving and which remain stuck.

Supporting Governance Conversations

Regular reviews of the decision inventory with leadership serve multiple purposes. They create accountability for whether key decisions are actually being made. They surface conflicts or ambiguities in decision authority. They prompt discussions about whether the decisions you're supporting are actually the most important ones.

The inventory provides a concrete artifact that makes decision quality a visible, managed concern rather than something assumed to happen automatically.

Maintaining Your Inventory Over Time

The decision inventory isn't a one time documentation exercise. It's a living component of your Strategy Layer that needs ongoing maintenance.

Add decisions as new strategic priorities emerge. When your organization launches a new initiative or shifts strategic focus, identify the key decisions that will drive success and add them to your inventory.

Remove or archive decisions that become obsolete. When a strategic goal is achieved, deprioritized, or delegated to automated systems, the associated decisions may no longer need systematic support. Archive them rather than maintaining them indefinitely.

Update ownership as roles change. When people move to new positions or organizational structure shifts, decision ownership may need to transfer. Keep the inventory current so it reflects actual authority and accountability.

Refine descriptions as understanding improves. Your initial documentation won't be perfect. As you work with decisions, you'll gain clarity about exactly what's being decided, what makes it difficult, and what information it needs. Update descriptions to reflect this learning.

Review quarterly or whenever strategy shifts significantly. Set a regular cadence for inventory review, typically quarterly, where you and relevant stakeholders assess whether the right decisions are documented, whether ownership is clear, and whether priorities should adjust.

Think of inventory maintenance like maintaining your project portfolio or asset register. It requires modest ongoing effort but pays dividends in organizational clarity and coordination.

Your Decision Inventory Template

Here's a simple template you can adapt for your first inventory:

Decision Name: [Clear statement of what's being decided]

Owner: [Role or person with final authority]

Frequency: [How often this occurs and typical timing]

Description: [2-3 sentences explaining the choice and why it matters]

Current Process: [Brief description of how this is done today and key pain points]

Stakeholders: [Who provides input or needs to be consulted]

Strategic Link: [Which strategic goal this influences and how]

Priority: [High/Medium/Low based on impact, frequency, and quality gap]

Maturity Assessment: [Current state: Ad hoc / Defined / Managed / Optimized]

Next Actions: [Immediate steps to improve this decision's support]

Start with 3 to 5 decisions using this template. Document them clearly. Share with decision owners and stakeholders. Use them to guide your next steps in building the Strategy Layer.

Implementation Exercise: Build Your First Decision Inventory

You've learned the concepts. Now build something real.

Time required: 2 to 3 hours for initial inventory

What you'll create: Documented inventory of 3 to 5 strategic decisions

Steps:

1. Choose your domain (15 minutes)

Select one strategic goal or business area. Write it at the top of a document: "Strategic Goal: [Your goal]"

2. Brainstorm decisions (30 minutes)

List 10 to 20 choices that influence whether you achieve that goal. Use the prompts from earlier in this chapter. Don't filter yet. Capture possibilities.

3. Prioritize to top 3 to 5 (30 minutes)

Rate candidates on strategic impact, frequency, current quality gap, and feasibility. Select your top 3 to 5. These are your starting inventory.

4. Document each decision (60 to 90 minutes)

For each prioritized decision, complete the template:

- Decision name and description
- Owner (propose if unclear)
- Frequency and timing
- Current process and pain points
- Stakeholders
- Strategic connection

- Priority and maturity rating

Be specific. Be honest. Capture reality, not aspiration.

5. Validate with one decision owner (15 minutes)

Share your documentation with the owner of at least one decision. Ask: Does this accurately reflect how this decision happens? Are the pain points right? Do you see value in making this more systematic?

Adjust based on their feedback.

What you have now: A working decision inventory that identifies your Strategy Layer starting point. These decisions will be the focus as you learn to design decision critical metrics (Chapter 4), create decision briefs (Chapter 5), and transform your analytics intake (Chapter 6).

Keep this inventory accessible. Reference it when prioritizing work. Update it as you learn. Use it as evidence that you're focusing analytics resources on decisions that matter.

You've made decisions visible. That's the foundation for everything that follows.

Moving Forward

With your decision inventory built, you've completed the first and most important step in creating your Strategy Layer. You've transformed invisible, informal decision making into documented, manageable components of your operational architecture.

But documentation alone doesn't improve decisions. The inventory is valuable only when you connect these decisions to the metrics that should inform them.

That's where Chapter 4 takes you next: designing decision critical metrics that directly serve the choices you've just cataloged.

Every metric you build from this point forward will justify its existence by answering the question: Which decision in my inventory does this support?

That simple discipline transforms measurement from a sprawling collection of interesting numbers into a focused system that drives better choices.

You're ready to design those metrics.

Chapter 4: Designing Decision-Critical Metrics

Organizations measure relentlessly. Customer satisfaction scores, employee engagement indices, project completion rates, revenue per employee, system uptime percentages, quality defect counts, inventory turns, time to fill for open positions. The list expands continuously.

Metrics accumulate in dashboards, spreadsheets, and slide decks. Leaders review them in meetings. Analysts refresh them on schedules. Yet when it comes time to make an important decision, those metrics often don't quite answer the questions that need answering.

The measurements exist, but they don't map cleanly to the choice at hand. So decisions get made anyway, informed by some combination of available metrics, informal data gathering, experience, and judgment.

This chapter transforms metrics from a long wishlist of things worth monitoring into a sharp, decision-tied set of measurements that directly enable the choices you inventoried in Chapter 3.

Three Types of Metrics

Not all metrics serve the same purpose. Understanding these differences prevents the common mistake of treating every measurement as equally important or expecting a single metric to serve multiple incompatible purposes.

Vanity Metrics

Vanity metrics look impressive but have minimal connection to decisions or actions. They often trend in desired directions simply because the business is operating, regardless of whether strategic choices are working.

Examples:

- Total number of customers without context about profitability or retention
- Total revenue without context about cost or sustainability
- Page views or downloads without context about engagement or conversion
- Social media follower counts without context about influence or business impact

These aren't useless. They might satisfy external stakeholders or provide broad context. But they rarely inform decisions directly. When a vanity metric moves, nothing changes about what the organization should do next.

Health Metrics

Health metrics monitor overall system performance and signal when something needs attention, but they don't specify what action to take.

Examples:

- System uptime percentages
- Overall employee turnover rate
- Aggregate customer satisfaction scores
- Total project count by status

They're valuable for awareness and for triggering deeper investigation when they move outside normal ranges. But they're diagnostic rather than prescriptive. If overall turnover increases, you need additional information to know whether the problem is compensation, management quality, workload, career development, or external market conditions.

Health metrics tell you something might be wrong. They don't tell you what to do about it.

Decision-Critical Metrics

Decision-critical metrics directly trigger or shape specific choices. They answer the exact questions that inform decisions with clear ownership and regular cadence.

Examples:

- Project schedule variance that determines whether to reallocate resources
- Customer churn risk scores that determine which accounts receive retention outreach
- Capacity utilization rates that determine whether to hire additional staff
- Inventory levels that trigger reordering decisions

These metrics have defined thresholds or ranges that map to different actions. When a decision-critical metric moves, someone knows what to do differently.

The same underlying measurement can serve different roles depending on context. Employee turnover might be a health metric for the CHRO monitoring overall workforce stability. It becomes decision-critical when it's segmented by role and location to inform specific hiring and retention investment decisions.

Revenue might be vanity when it's just a total number celebrated in company meetings. It becomes decision-critical when it's measured by segment and compared to acquisition cost to inform which customer types to prioritize.

The Strategy Layer focuses primarily on decision-critical metrics. Health metrics have their place in monitoring and early warning. Vanity metrics might serve external communications. But neither should be confused with metrics that actually drive decisions.

What Makes a Metric Decision-Critical

Several characteristics distinguish metrics that enable decisions from metrics that simply describe situations.

Linked to a Specific Decision

A decision-critical metric must connect explicitly to a decision in your inventory. Someone should be able to explain which decision this metric informs, who owns that decision, and how often it's made.

If you can't articulate this connection, the metric may be valuable for other purposes but isn't decision-critical.

Test: For any metric you're considering, ask "Which decision in our inventory does this support?" If the answer is vague or requires explanation, it's not decision-critical yet.

Has Clear Direction

It must be obvious whether higher or lower values are better, or whether the metric should stay within a specific range.

Ambiguous metrics that could be interpreted multiple ways don't enable decisions. They create debates about interpretation instead.

Clear examples:

- Net retention rate: higher is better
- Customer acquisition cost relative to lifetime value: lower ratios are better

- System response time: lower is better, with acceptable range defined

Ambiguous example:

- Number of strategic initiatives without context about capacity or completion: Is more better because it shows ambition? Or is less better because it shows focus?

Has Known Thresholds

Decision-critical metrics need defined levels that trigger different actions. This doesn't necessarily mean hard cutoffs, but there should be ranges everyone understands.

Examples:

- If project schedule variance exceeds 15%, we escalate to executive attention
- If customer health scores drop below 60, we trigger retention outreach
- If critical position time to fill exceeds 90 days, we expand recruiting sources or adjust requirements

Without thresholds, metrics inform general awareness but don't directly trigger decisions.

Has Accountable Owner

Someone must be accountable for both the metric's accuracy and for acting on what it shows. This connects to decision ownership. The person who owns the decision should own or explicitly rely on the metrics that inform it.

Metrics without owners become orphaned over time as definitions drift, data sources change, or organizational memory fades about why the metric mattered.

Reviewed at Decision Cadence

Decision-critical metrics should be examined with the same frequency as the decisions they inform.

If you make portfolio funding decisions quarterly, the metrics that inform those decisions should be reviewed quarterly in preparation. If you make staffing decisions weekly, the relevant metrics need weekly updates.

Metrics calculated frequently but reviewed rarely fall out of sync with decisions. Metrics reviewed frequently but updated slowly create false precision.

The Decision to Metric Design Process

Here's how to design decision-critical metrics for the decisions in your inventory:

Step 1: Start With the Decision

Never design metrics in isolation. Always begin with a specific decision from your inventory.

Pull up one decision. Read the description. Understand who owns it, how often it occurs, what makes it difficult, and what strategic goal it serves.

Example decision from inventory: "Decide which at-risk enterprise customers to target for retention programs each quarter"

- Owner: VP of Customer Success
- Frequency: Quarterly
- Current pain: No systematic way to identify highest risk accounts or predict which interventions work

Step 2: Identify Essential Questions

What do you need to know to make this decision well? List the questions that, if answered, would make the choice clearer.

For the retention decision above:

Essential questions:

- Which customers are at highest risk of churning?
- Among at-risk customers, which represent highest value to retain?
- What factors are driving churn risk for different customer segments?
- Which retention tactics have worked historically for similar situations?
- What's our capacity to execute retention programs this quarter?

These questions bridge between the decision and the metrics. They translate "decide what to do" into "understand these specific things."

Step 3: Design Metrics That Answer Questions

For each essential question, design one or two metrics that provide a clear answer.

Question: Which customers are at highest risk of churning? **Metric:** Customer churn risk score (0-100 scale based on usage decline, support tickets, contract renewal timing, and engagement patterns) **Threshold:** Scores above 70 trigger retention review

Question: Among at-risk customers, which represent highest value? **Metric:** Annual recurring revenue at risk (total ARR for customers with churn risk score above 70) **Threshold:** Prioritize customers representing more than $100K ARR

Question: What factors drive churn risk? **Metric:** Primary churn driver by customer segment (categorized as: product gaps, service issues, pricing concerns, competitive pressure, organizational change) **Usage:** Informs which retention tactics to deploy

Question: Which retention tactics work historically? **Metric:** Retention program success rate by intervention type and customer segment **Usage:** Guides program selection based on situation match

Question: What's our execution capacity? **Metric:** Available customer success manager hours for retention work this quarter **Threshold:** Determines how many accounts we can realistically target

Notice how each metric directly answers a specific question needed for the decision. They're not generic measurements of customer health. They're focused on what the VP of Customer Success needs to know to decide which retention programs to fund this quarter.

Step 4: Specify Metric Details

For each metric, document the specifics that make it calculable and actionable:

Metric name: Customer churn risk score

Definition: Composite score (0-100) predicting likelihood of non-renewal based on four weighted factors: usage trend (40%), support ticket velocity (25%), engagement with key features (20%), time since last executive contact (15%)

Calculation method: Automated daily calculation using CRM and product usage data, with scores normalized across customer base

Data sources:

- Product analytics platform for usage data
- Support ticketing system for ticket volume and resolution
- CRM for customer engagement tracking

Update frequency: Daily calculation, weekly review by customer success team, comprehensive quarterly review before retention planning

Owner: Director of Customer Analytics (calculation), VP of Customer Success (action on results)

Thresholds:

- 0-40: Low risk, standard engagement
- 41-70: Moderate risk, monitor closely
- 71-85: High risk, retention review recommended
- 86-100: Critical risk, immediate executive engagement

This level of detail ensures the metric can be built consistently, maintained reliably, and interpreted correctly by everyone who uses it.

Step 5: Validate With Decision Owner

Before finalizing metrics, review them with the person who will use them to make decisions.

Questions to ask:

- Do these metrics answer the questions you need answered?
- Are the thresholds set at levels that make sense for how you'll decide?
- Is the update frequency aligned with when you make this decision?
- Are there critical questions we're not addressing?
- Are any of these metrics measuring things that don't actually matter for your decision?

Decision owners often reveal gaps or unnecessary complexity at this stage. Better to discover this before investing in building metrics than after they're deployed.

Common Design Pitfalls

Watch for these mistakes when designing decision-critical metrics:

Pitfall 1: Metrics That Don't Match Decision Frequency

You design a metric that updates monthly to support a decision made weekly. By the time fresh data arrives, the decision has already been made multiple times using stale information.

Fix: Match metric update cadence to decision cadence. If that's not feasible with current data sources, either adjust decision timing or find proxy metrics available at the right frequency.

Pitfall 2: Thresholds Set Without Decision Context

You define a metric but leave thresholds vague: "we want customer satisfaction to be high" or "keep costs reasonable."

Fix: Work with decision owners to specify exact threshold values: "customer satisfaction below 75 triggers investigation" or "cost variance exceeding 10% requires approval for continued spending."

Pitfall 3: Metrics That Measure Outputs Instead of Outcomes

You track activities rather than results. Number of customer calls made instead of customer retention rate. Number of training sessions delivered instead of skill improvement.

Fix: Focus metrics on outcomes that matter for the decision. Activities might be useful for implementation tracking, but decision-critical metrics should measure results.

Pitfall 4: Overly Complex Composite Metrics

You create sophisticated indices that combine many factors in ways that become opaque. When the metric moves, no one can diagnose why or what to do about it.

Fix: Keep metrics as simple as possible while still answering the question. If complexity is necessary, ensure you can decompose the metric to understand drivers.

Pitfall 5: Metrics Divorced From Actions

You design elegant measurements that provide insight but don't connect to specific actions the decision owner can take.

Fix: Always complete this sentence: "When this metric shows X, the decision owner will do Y." If you can't complete it, the metric isn't decision-critical yet.

Metric Design Examples Across Decision Types

Here's how decision-critical metrics look for different types of decisions:

Investment Decision: IT Portfolio Funding

Decision: Which projects to continue funding versus defer or cancel at quarterly portfolio reviews

Essential questions:

- Which projects are behind schedule or over budget?
- Which projects still align with current strategic priorities?
- Which projects are blocking other high-value work?
- What's the opportunity cost of continuing versus stopping?

Decision-critical metrics:

Schedule variance by project

- Definition: Percentage deviation from planned completion date
- Threshold: More than 20% behind triggers review for cancellation
- Source: Project management system
- Owner: PMO Director

Strategic alignment score

- Definition: Weighted assessment (1-5 scale) of project contribution to current strategic priorities
- Threshold: Score below 3 makes project candidate for deferral
- Source: Quarterly strategy review with executives
- Owner: VP of Strategy

Dependency impact index

- Definition: Count of high-priority projects waiting on this project's completion
- Threshold: Index above 3 elevates project priority regardless of other metrics
- Source: Project dependency mapping
- Owner: Portfolio Manager

Operational Decision: Workforce Allocation

Decision: How to allocate nursing staff across hospital units and shifts based on daily census

Essential questions:

- What's current patient load by unit?

- How complex are patient needs in each unit?
- What staffing changes are anticipated in next 24 hours?
- Which units are most under-resourced relative to need?

Decision-critical metrics:

Census by unit

- Definition: Current occupied beds as percentage of capacity
- Threshold: Above 90% triggers staffing increase consideration
- Source: Patient management system (real-time)
- Owner: Shift supervisors

Acuity-adjusted staffing ratio

- Definition: Nurses per patient adjusted for patient complexity scores
- Threshold: Ratio below 1:5 for high-acuity patients triggers immediate review
- Source: Clinical documentation system
- Owner: Nursing operations manager

Anticipated admission/discharge volume

- Definition: Expected patient movements in next 12 hours based on historical patterns and current pipeline
- Threshold: Expected 20%+ census change triggers proactive staffing adjustment
- Source: Admission scheduling system plus historical patterns
- Owner: Capacity planning team

Risk Decision: Security Incident Response

Decision: Which security alerts to investigate immediately versus monitoring over time

Essential questions:

- How severe would impact be if this threat is real?
- How confident are we this is a true positive versus false alarm?
- What assets or systems are potentially affected?
- Do we have capacity to investigate now?

Decision-critical metrics:

Threat severity score

- Definition: Potential impact rating (1-10) based on affected systems, data sensitivity, and attack sophistication
- Threshold: Score 7+ requires immediate investigation
- Source: Security information and event management system
- Owner: Security operations center manager

Alert confidence level

- Definition: Probability (percentage) that alert represents genuine threat based on signature matching and behavior analysis
- Threshold: Confidence above 60% combined with severity 7+ triggers immediate action
- Source: Threat detection algorithms
- Owner: Senior security analyst

Response capacity indicator

- Definition: Available analyst hours this shift versus active investigations
- Usage: Determines whether to investigate now or queue for next shift
- Source: Incident tracking system
- Owner: SOC shift lead

Notice how each set of metrics is tailored to the specific decision. They're not generic health monitoring. They directly answer questions needed to make that particular choice at that frequency.

Connecting Metrics to Strategy

Every decision-critical metric should trace back through decisions to strategic goals. This connection ensures you're measuring what matters, not just what's easy to measure.

Create a simple map:

Strategic Goal → Critical Decision → Essential Question → Decision-Critical Metric

Example:

Goal: Increase recurring revenue 20% annually | **Decision:** Which at-risk customers to target for retention | **Question:** Which customers are highest risk? | **Metric:** Customer churn risk score with 70+ threshold

Goal: Accelerate project delivery 25% | **Decision:** Which projects to continue versus defer | **Question:** Which projects are furthest behind schedule? | **Metric:** Schedule variance percentage by project

Goal: Improve operational safety to top quartile | **Decision:** Where to invest in safety improvements | **Question:** Which locations have highest incident rates? | **Metric:** Recordable incidents per 100,000 hours worked by site

This mapping serves multiple purposes. It shows executives how metrics connect to strategy. It helps analytics teams prioritize metric maintenance. It reveals gaps where strategic goals lack adequate measurement. It provides rationale for retiring metrics that don't trace back to strategy.

Metric Governance Essentials

Decision-critical metrics need governance to maintain quality and relevance over time.

Metric Definition Documentation

Every decision-critical metric needs a documented specification that includes:

- Metric name and purpose (which decision it serves)
- Precise calculation method
- Data sources and systems
- Update frequency
- Ownership (who's accountable for accuracy)
- Thresholds and action triggers
- Historical baseline and targets

This documentation prevents the common problem where "customer satisfaction" means different things to different people, or where calculation methods drift over time making historical comparisons meaningless.

Change Control Process

Changes to decision-critical metrics should happen deliberately, not casually. When someone proposes modifying a metric definition, calculation, or threshold, require:

- Rationale for the change

- Impact assessment (how will this affect historical trends and current decisions)
- Review by decision owner and key stakeholders
- Documentation of what changed and when
- Communication to everyone who uses the metric

Minor clarifications might be approved by metric owner and decision owner. Significant changes affecting thresholds or calculation methods need broader review.

Metric Retirement

When decisions change or metrics prove ineffective, retire them rather than maintaining indefinitely.

Triggers for retirement:

- Decision no longer in inventory (goal achieved, deprioritized, or automated)
- Metric consistently unused in decision process (logged but not referenced)
- Metric doesn't correlate with outcomes (predicted and actual results diverge)
- Better alternative metric developed

Retirement means documentation is updated to show the metric is inactive, calculation stops, and the metric disappears from reports. Archived historical data remains for reference but isn't maintained going forward.

A regional bank implemented this lifecycle and retired 30% of their metrics in the first year. These weren't randomly chosen. Analysis revealed they weren't being used in decisions. Some dated to requirements from executives who'd left years ago. Some measured things that seemed important but turned out not to correlate with outcomes anyone cared about.

The result wasn't diminished capability. It was focus. The analytics team spent less time maintaining unused metrics and more time improving metrics that mattered. Data quality improved because they could invest in fewer, more important measurements.

Your Metric Design Template

Use this template to design decision-critical metrics for decisions in your inventory:

For Decision: [Name from your decision inventory]

Essential Question: [What you need to know to decide well]

Metric Name: [Clear, descriptive name]

Purpose: [Which specific decision this informs and how]

Definition: [Precise specification of what's being measured]

Calculation Method: [Exactly how the metric is computed]

Data Sources: [Systems or processes providing the data]

Update Frequency: [How often calculated and reviewed]

Owner: [Who's accountable for accuracy and action]

Thresholds:

- [Level 1]: [What this means and what action follows]
- [Level 2]: [What this means and what action follows]
- [Level 3]: [What this means and what action follows]

Validation: [How decision owner confirmed this metric serves their needs]

Related Metrics: [Other metrics that provide context or work together]

Complete this template for each metric supporting the 3-5 decisions in your initial inventory. Don't design metrics that don't connect to documented decisions.

Implementation Exercise: Design Your First Decision-Critical Metrics

Time required: 2-3 hours

What you'll create: Full metric specifications for one decision from your inventory

Steps:

1. Select one decision (5 minutes)

Choose one decision from your Chapter 3 inventory. Pick something important where you have access to decision owner and relevant data.

2. List essential questions (15 minutes)

What does the decision owner need to know to make this choice well? Write 3-5 questions. Be specific. Not "what's customer health?" but "which customers are at highest risk of churning in next 90 days?"

3. Design metrics for each question (60 minutes)

For each question, design 1-2 metrics that provide clear answers. Use the template above. Include:

- Definition specific enough to calculate
- Data sources you can realistically access
- Thresholds that map to actual decisions

- Update frequency matching decision cadence

Don't worry about perfect. Focus on usable.

4. Validate with decision owner (30 minutes)

Schedule 30 minutes with the person who makes this decision. Walk through your proposed metrics. Ask:

- Do these answer the questions you need answered?
- Are thresholds set at sensible levels?
- Is anything critical missing?
- Is anything here that doesn't actually matter?

Adjust based on feedback.

5. Document final specifications (30 minutes)

Complete the template for each validated metric. Include all details: definition, calculation, sources, ownership, thresholds.

What you have now: Decision-critical metrics ready to build and deploy. These connect directly to a specific decision in your inventory with validation from the person who'll use them.

Next step: Chapter 5 shows you how to package these metrics into decision briefs that executives actually use to make choices.

Moving Forward

You've now completed two foundational steps in building your Strategy Layer:

Chapter 3: Built your decision inventory (documented important choic-es)**Chapter 4:** Designed decision-critical metrics (measurements that inform those choices)

The connection between decisions and metrics is now explicit. Every metric justifies its existence by serving a documented decision. Every decision has clarity about what information it needs.

But having good metrics doesn't guarantee they'll influence decisions. That requires structuring how information gets presented to decision makers.

Chapter 5 teaches you to create decision briefs: the format that packages your metrics and analysis into presentations that drive action instead of generating more meetings.

The gap between having information and using information to decide is what decision briefs close.

You're ready to learn how.

Chapter 5: Decision Briefs That Drive Action

The meeting follows a familiar pattern. An analyst presents carefully prepared slides dense with charts, tables, and insights. The executive listens for a few minutes, interrupts with questions the analysis didn't anticipate, and the meeting ends with vague agreement to "look at this more" or "circle back next month."

Everyone leaves frustrated. The analyst spent days building something that didn't land. The executive still doesn't have what's needed to decide. The decision gets deferred again.

This pattern repeats because most analytic work isn't structured for how executives actually make decisions. It's designed for how analysts think decisions should be made: with comprehensive data, rigorous methodology, and patient examination of nuance.

This chapter shows you how to present analysis in the format executives actually need: decision briefs that frame choices clearly, present evidence concisely, and end with action instead of more discussion.

What Executives Actually Need

Understanding what executives need starts with understanding their context. They're accountable for outcomes they can't personally deliver. They face more decisions than time allows. They work with incomplete information and competing pressures.

In this reality, executives don't need comprehensive background or methodological detail. They need something specific.

Options and Trade-offs

Executives rarely need to know what the data says in abstract terms. They need to understand their realistic choices and how those choices differ in consequences.

Not "customer churn increased 3% last quarter" but "we can invest $2M in retention programs targeting high value segments, accept the churn and redirect resources to acquisition, or defer the decision pending more analysis. Here's what each option likely produces."

The data matters primarily as evidence for evaluating options, not as interesting facts to contemplate.

Forward-Looking Implications

Dashboards typically show what happened. Last quarter's revenue, last month's attrition, yesterday's system performance. Executives need this context, but what they really need is implications.

If this trend continues, what happens? If we make this choice, what do we expect to change? What's at stake if we delay deciding?

Historical data serves decision making only when it's translated into forward-looking scenarios.

Confidence About Strategic Alignment

When executives review metrics, they need assurance that these measurements actually matter for strategic goals. Too often, they're presented with whatever metrics happen to be available or whatever analysts thought was interesting, without clear connection to whether the organization is winning or losing at what matters most.

The Strategy Layer provides this confidence by ensuring every metric traces back through decisions to strategic outcomes.

Clarity About What's Being Asked

Many presentations leave executives uncertain what they're supposed to do with the information. Is this for awareness? For approval? For decision?

If for decision, what's the decision, when must it be made, and what authority or resources are required?

Executives need explicit requests: "We need you to choose between Option A and Option B by end of month" or "We need you to approve additional budget for X based on this evidence."

Recommendations With Clear Points of View

Executives value objectivity, but they don't value false neutrality. When an analyst has examined the data, understands the context, and has relevant expertise, executives expect that analyst to take a position.

Not "here's what the data shows, you decide" but "based on this evidence, my recommendation is X because of Y and Z, though there are risks including W that you should consider."

Executives can disagree with recommendations, but they shouldn't have to reverse engineer what the analyst thinks should happen.

The Decision Brief Format

The decision brief is a structured, typically one-page format that gives executives exactly what they need. It frames a specific decision with all essential information for choosing wisely.

Element 1: Context (What's at Stake)

Begin with strategic importance and urgency. Why does this decision matter? What happens if we don't decide, or decide poorly? What's the broader situation this decision sits within?

Context shouldn't be lengthy background. Two to three sentences that orient the executive to why this matters now.

Example: "Customer retention has declined 8% in our enterprise segment over six months, putting $12M in annual recurring revenue at risk. We need to decide on retention investment before Q4 planning closes in three weeks. Without intervention, projected losses could reach $20M annually by year end."

This tells the executive immediately what's important (retention decline), what's at stake ($12M-$20M), and what's urgent (three week deadline).

Element 2: Decision (What Needs to Be Decided Now)

State the explicit choice being made, framed as a clear question with bounded scope.

Not "what should our retention strategy be" but "should we invest $2M in a targeted enterprise retention program for Q4, or maintain current approach?"

The decision statement makes clear what's being decided and what's not being decided in this moment. It prevents scope creep where a focused decision expands into a strategic review that delays action.

Example: "Should we invest $2M in targeted retention programs for at-risk enterprise accounts, focusing on the top 50 accounts representing $8M in ARR, or should we maintain our current reactive approach and reallocate these funds to new customer acquisition?"

This is bounded (targeted program versus current approach, not comprehensive strategy revision), specific ($2M, top 50 accounts), and decidable (clear alternatives).

Element 3: Options (Realistic, Clearly Contrasted)

Present two to four realistic alternatives. Not every possible variation, but the genuinely viable choices.

For each option, describe what would actually happen (actions, resources, timeline) and expected outcomes based on available evidence. Contrast options on the dimensions that matter most for this decision: typically cost, timing, risk, and expected impact.

Avoid false choices or straw man options included just to make the preferred option look better. Executives see through this and it undermines credibility.

Example:

Option A: Targeted Retention Program ($2M investment)

- Deploy dedicated customer success resources to top 50 at-risk accounts
- Implement personalized engagement plans based on churn risk drivers
- Offer strategic account reviews and product roadmap input
- Timeline: Launch in 30 days, run through Q4
- Expected outcome: Retain 70% of targeted accounts (historical program success rate), saving $5.6M in ARR
- Risk: Investment may not prevent churn if issues are primarily product related rather than engagement related

Option B: Maintain Current Approach ($0 additional investment)

- Continue reactive retention efforts when customers signal intent to leave
- Reallocate $2M to new customer acquisition programs
- Timeline: Immediate

- Expected outcome: Likely lose 60-70% of at-risk accounts ($4.8M-$5.6M in ARR), but acquire estimated 50 new customers generating $3M in new ARR
- Risk: Net revenue decline of $1.8M-$2.6M, plus signal to market that we don't prioritize existing customers

The contrast is clear. Option A costs money upfront but protects existing revenue. Option B avoids the cost but accepts losses while chasing new revenue. Neither is perfect. The executive can see the trade-off.

Element 4: Evidence (Decision-Critical Metrics)

Show the specific measurements from your Strategy Layer that inform this choice. Not a comprehensive data dump, but the three to five metrics that most directly address whether each option would work.

Include actual numbers with context (current state, trend, comparison to benchmarks or targets) and clear interpretation of what the metrics mean for the options being considered.

If data is incomplete or uncertain, say so explicitly rather than hiding limitations.

Example:

Churn Risk Distribution:

- 50 enterprise accounts currently score above 70 on churn risk scale (high risk threshold)
- These accounts represent $8M in ARR (18% of enterprise segment revenue)
- Primary churn drivers: 40% product feature gaps, 35% service responsiveness, 25% pricing pressure

Historical Program Performance:

- Previous targeted retention programs achieved 70% success rate (35 of

50 customers retained)

- Programs cost average $40K per retained customer
- Payback period: 6 months on average

Acquisition Alternatives:

- $2M in acquisition spending typically yields 50 new enterprise customers
- Average new customer ARR: $60K (versus $160K average for at-risk accounts)
- New customer acquisition payback period: 18 months on average

Competitive Context:

- Industry benchmark retention rate: 92% (we're at 87% currently)
- Competitors increasing retention investment by average 25% this year

These metrics directly inform the choice. They show what's at risk, whether retention programs work, how retention economics compare to acquisition economics, and how we're performing relative to competition.

Element 5: Recommendation (A Clear Point of View)

State which option you recommend and why, based on strategic priorities, risk tolerance, and the evidence presented.

Acknowledge trade-offs and risks in your recommended approach. No option is perfect. If there are conditions under which a different option would be better, specify them.

Take a position and defend it. The recommendation shouldn't be hedged or diplomatic to the point of meaninglessness.

Example:

"Recommend Option A: Invest $2M in targeted retention program. Rationale: Protecting existing high-value customers aligns with our strategic priority of increasing customer lifetime value and produces better economics than acquisition ($40K retention cost versus $60K acquisition cost per customer, with 3x faster payback). Risk that 30% of at-risk customers churn anyway is acceptable given potential to save $5.6M in ARR. Alternative Option B would produce net revenue decline of approximately $2M while signaling to market that we don't invest in customer success."

The recommendation is clear. The reasoning is explicit. The risks are acknowledged. The executive knows what you think should happen and why.

Structure Matters More Than Length

Decision briefs can be one-page documents, short slide decks, or structured verbal presentations. What matters isn't the format but the discipline of addressing each element clearly.

Some executives prefer to read briefs 24-48 hours before meetings and use meeting time for discussion. Others prefer verbal presentation of the brief structure during the meeting. Adapt the delivery to executive style, but maintain the five-element structure.

One-page brief: Dense but readable, all five elements visible at once

Slide deck: One slide per element, 5 total core slides plus appendix for supporting detail

Verbal presentation: Walk through the five elements in order, 10-15 minutes total, followed by discussion

What doesn't work is burying the decision in slide 15 after comprehensive background, or presenting findings without connecting them to a choice someone needs to make.

Before and After: A Transformation Example

Here's how the same analysis looks presented traditionally versus as a decision brief:

Traditional Approach (Doesn't Work)

Slide 1: Title and agenda **Slides 2-8:** Background on enterprise customer segment, market trends, competitive landscape **Slides 9-15:** Detailed churn analysis by cohort, geography, product line, and customer characteristic **Slides 16-20:** Regression analysis showing factors correlated with churn **Slides 21-25:** Case studies of individual customer situations **Slide 26:** Summary of findings **Slide 27:** Questions for discussion

Total: 27 slides, 45 minute presentation, ending with "what should we do about this?"

The executive has absorbed lots of information but isn't closer to making a decision. What are the actual options? What do you recommend? When does this need to be decided?

Decision Brief Approach (Works)

Context: Enterprise churn increased 8%, putting $12M ARR at risk. Decision needed in 3 weeks before Q4 planning closes.

Decision: Invest $2M in targeted retention program or maintain current approach?

Options:

- Option A: Targeted program for top 50 accounts, 70% expected success rate, $5.6M ARR saved
- Option B: Accept churn, invest $2M in acquisition instead, net $2M revenue decline

Evidence:

- 50 high-risk accounts representing $8M ARR
- Historical retention programs: 70% success rate, $40K cost per retained customer
- Acquisition alternative: 50 new customers at $60K ARR each, 18-month payback versus 6-month retention payback

Recommendation: Option A. Better economics, faster payback, strategic alignment with customer lifetime value goals.

Total: 1 page or 5 slides, 10 minute presentation, executive ready to decide.

All the background from the traditional approach still exists. It's available in appendix if the executive wants to dig deeper. But the decision-critical information is front and center, structured to enable choice.

Common Mistakes and How to Fix Them

Mistake 1: Burying the Decision

The analyst presents extensively on background and analysis before finally getting to "so, we need to decide whether to do X or Y" on slide 20.

By this point, the executive has lost the thread. The context for why this matters is buried in slides that were presented 15 minutes ago.

Fix: Lead with the decision. State it clearly in the first 30 seconds. "Today we need to decide whether to invest in retention or redirect to acquisition. Let me give you the context, options, and evidence."

Mistake 2: Options That Aren't Really Options

The analyst presents three options but two of them are obviously non-viable, included just to show they were considered.

Fix: Only present options you believe are genuinely reasonable. If there's one clear right answer, say so and use the brief to build the case. Don't pad with fake options to create the appearance of balance.

Mistake 3: Data Without Interpretation

The analyst shows charts and tables but doesn't explicitly connect them to what the executive should conclude.

"Here's the churn rate by customer segment" (chart displayed) followed by silence, expecting the executive to draw implications.

Fix: Always interpret your data. "This chart shows churn is concentrated in mid-market accounts, which means a retention program should focus there rather than spreading resources across all segments."

Mistake 4: Hedged Recommendations

The analyst concludes with "both options have merit" or "it depends on our priorities" without taking a clear position.

Fix: Make a call. Say which option you recommend and why. You might be wrong. The executive might choose differently. But your job is to provide informed judgment, not to avoid accountability by staying neutral.

Mistake 5: Ignoring the "So What"

The analyst presents facts without connecting them to action: "Customer satisfaction declined from 78 to 74."

The executive's internal response: "Okay, so what should we do about that?"

Fix: Always complete the thought. "Customer satisfaction declined from 78 to 74, which historically leads to 10% churn increase. We should decide whether to invest in service improvements now or accept higher attrition."

Adapting to Different Executive Styles

Executives vary in how they prefer to receive and process information, but the five-element structure adapts to different styles.

The Reader

Wants full brief in advance, reads carefully, comes to meeting with questions.

Adaptation: Provide one-page brief or slide deck 24-48 hours early. Keep meeting short, focused on questions and decision. This executive has usually formed a view and uses the meeting to test it.

The Discusser

Wants to think out loud, debate options, and arrive at decisions through conversation.

Adaptation: Present brief verbally but move quickly through context and evidence to get to options and discussion. These executives need space to explore implications and challenge assumptions. Expect the meeting to be longer but result in clear decision.

The Visual Processor

Grasps complex situations quickly through well-designed graphics but loses attention during text-heavy presentations.

Adaptation: Lead with visual comparison of options. Use charts that directly contrast alternatives. Minimize text. Keep verbal explanation concise.

The Detail Seeker

Wants to understand methodology, probe assumptions, and examine edge cases.

Adaptation: Have backup material ready. Present the brief structure to frame the decision, then be prepared to dive into specifics when asked. These executives need to understand the foundation before they'll commit to the conclusion.

What doesn't change regardless of style is the discipline of framing a clear decision, presenting realistic options, providing decision-critical evidence, and offering a recommendation.

Integrating With Your Strategy Layer

Decision briefs become powerful when they're integrated with the Strategy Layer components you've built in previous chapters.

Drawing From Your Decision Inventory

The decision you're framing should be in your inventory from Chapter 3. You already know who owns it, how often it occurs, what makes it difficult, and what strategic goal it serves.

This context ensures you're focusing on decisions that matter, not just interesting analysis.

Using Your Decision-Critical Metrics

The evidence section draws directly from metrics you designed in Chapter 4. These aren't random measurements but specific metrics created to inform this exact decision.

Because metrics and decisions are explicitly connected, you know you're showing information that actually matters for the choice.

Enabling Systematic Improvement

When you structure analysis as decision briefs that follow from your Strategy Layer, several things improve systematically:

Requests become clearer. Before asking for analysis, requesters must frame the decision. This clarity reduces misunderstanding and rework.

Analytics work focuses better. Analysts know exactly which decision they're supporting and can structure work accordingly from the start.

Executives get consistent format. Over time, executives become accustomed to the brief structure and can absorb information faster.

Learning compounds. When you consistently document what decision was made, what evidence informed it, and what outcomes resulted, you build organizational knowledge about what works.

Your Decision Brief Template

Use this template to structure your next presentation:

DECISION BRIEF

Date: [When this brief was prepared] **For:** [Decision owner and stakeholders] **Prepared by:** [Your name and role]

1. CONTEXT (2-3 sentences)

- What strategic issue or opportunity are we addressing?
- What's at stake?
- Why does this need to be decided now?

2. DECISION (1-2 sentences)

- State the explicit choice being made
- Clarify what's in scope and what's not

3. OPTIONS (2-4 alternatives)

For each option:

- **Description:** What would we actually do?
- **Expected outcome:** What results do we anticipate?
- **Key trade-offs:** What are we giving up or risking?
- **Timeline and cost:** When and how much?

4. EVIDENCE (3-5 decision-critical metrics)

For each key metric:

- **Current value and trend**
- **What this tells us about options**
- **Data confidence level**

5. RECOMMENDATION

- Which option do you recommend?
- Primary rationale
- Key risks acknowledged
- Conditions that might change recommendation

APPENDIX (if needed)

- Supporting analysis
- Methodology details
- Alternative scenarios examined

- Implementation considerations

Implementation Exercise: Transform One Analysis Into a Decision Brief

Time required: 90 minutes

What you'll create: A complete decision brief for a real upcoming decision

Steps:

1. Choose your decision (10 minutes)

Select one decision from your Chapter 3 inventory that you need to support in the next 2-4 weeks. Something where analysis is needed and the executive must choose between real alternatives.

2. Draft the five elements (60 minutes)

Context (10 minutes): Write 2-3 sentences explaining what's at stake and why this matters now. Test: Would someone unfamiliar with your organization understand why this is urgent?

Decision (5 minutes): Frame the explicit choice. Make it specific enough to decide, bounded enough to avoid scope creep. Test: Can you state this as a clear question with defined alternatives?

Options (25 minutes): Describe 2-3 realistic alternatives. For each, specify actions, expected outcomes, and key trade-offs. Test: Would the decision owner recognize these as the actual options being considered?

Evidence (15 minutes): Pull 3-5 metrics from your Chapter 4 work that directly inform this choice. Show current values, trends, and what they mean for evaluating options. Test: Does each metric clearly support comparing the options?

Recommendation (5 minutes): State which option you recommend and why. Acknowledge main risks. Be clear and definitive. Test: Would someone reading this know exactly what you think should happen?

3. Get feedback (20 minutes)

Share your draft brief with a colleague or the decision owner. Ask:

- Is the decision framed correctly?
- Are these the right options?
- Does the evidence help you evaluate choices?
- Is the recommendation clear?

Adjust based on what you hear.

What you have now: A presentation structure you can use for your next executive meeting. Instead of building a comprehensive deck that might or might not land, you have a focused brief that frames a clear choice.

Next step: Use this brief in the actual decision meeting. Note what works and what gets questioned. Refine your approach based on real feedback.

Moving Forward

You've now built three core components of your Strategy Layer:

Chapter 3: Decision inventory documenting your most important choices **Chapter 4:** Decision-critical metrics designed to inform those choices **Chapter 5:** Decision brief format for presenting choices and evidence

These pieces work together. Decisions come from your inventory. Metrics supporting those decisions come from your metric map. Briefs structure how you present metrics to decision owners.

But even with great briefs, you'll still face a flood of analytics requests that don't connect to clear decisions. Chapter 6 shows you how to transform your analytics intake process so that decision context becomes required before work begins.

That transformation shifts analytics from reactive report factory to strategic decision partner.

You're ready to redesign how work enters your analytics function.

Chapter 6: Transforming Analytics Intake

Your analytics team receives more requests than it can handle. Someone needs "some numbers on customer behavior." A manager wants "a quick report on project status." An executive asks for "analysis of our competitive position."

The requests arrive via email, Slack messages, hallway conversations, and ticketing systems. Each one feels urgent to the requester. Most are vaguely defined. Few specify what decision they're meant to inform.

Your team does what it can. Analysts interpret vague requests, build what seems reasonable, and hope it lands. Sometimes it does. Often it doesn't. Either way, more requests arrive tomorrow, and the cycle continues.

This chapter transforms that reactive, overwhelming intake process into a disciplined system where decision context is required before work begins. The result: analytics capacity flows to decisions that matter, requesters get better support, and your team stops being a report factory.

The Problem With Current Intake

Most analytics teams operate with some form of request intake, whether formal ticketing systems or informal email inboxes. The typical pattern creates several problems.

Vague Requests Waste Time

"Can you pull customer satisfaction scores by region?" seems straightforward. An analyst might spend hours creating a comprehensive report with multiple visualizations, regional comparisons, trend analysis, and demographic breakdowns, only to discover the requester just needed a quick number for a presentation slide.

Without understanding the actual need, analysts either overbuild (wasting time on unused detail) or underbuild (requiring rework when it turns out more was needed).

Prioritization Happens Through Politics

When everything feels urgent and nothing connects to documented decisions, prioritization defaults to organizational politics. Requests from senior executives jump the queue regardless of strategic value. Persistent requesters who follow up aggressively get faster service. Work gets done in the order it arrives, regardless of importance.

None of these prioritization methods correlate with decision impact or strategic value.

Rework Cycles Consume Capacity

Initial requests are rarely right. The requester didn't fully think through what they needed. The analyst misunderstood the context. What gets delivered doesn't quite match the actual need.

So the work cycles back: "This is good, but can you also show it by product line?" Then: "Actually, we need last year's numbers for comparison." Then: "The executive wants to see this at the customer segment level."

Each revision consumes capacity that could have been used for other work. The underlying problem is that neither requester nor analyst clarified the decision context at the beginning.

Important Work Gets Delayed

While your team responds to whoever asks loudest, critical decisions that would benefit from systematic support go unsupported. The quarterly portfolio funding decision that affects $50M in investment happens without adequate analysis because analysts are busy building reports that don't connect to clear decisions.

The work that matters most gets crowded out by the work that's most visible or most persistent.

Analysts Feel Like Order Takers

Your team members became analysts because they wanted to solve problems and influence decisions. Instead, they spend most of their time responding to vague requests, guessing at what's really needed, and watching their work get used briefly or not at all.

This frustration leads to turnover, burnout, and difficulty attracting strong talent. Good analysts want their work to matter.

The Transformation: Decision Context Required

The solution is straightforward but requires discipline. Every analytics request must specify the decision it's meant to inform before work begins.

Not as an aspirational practice for important requests. As a requirement for all requests.

What This Means in Practice

When someone asks for analytics work, they must answer:

- **Which decision will this inform?** (Ideally from your Chapter 3 inventory)
- **Who owns that decision?** (The person with authority to choose)

- **When will the decision be made?** (Specific timing, not "eventually")
- **What options are being considered?** (The actual alternatives)
- **What would change based on different results?** (How the analysis affects the choice)

If a requester can't answer these questions, the request isn't ready for analytics work. It might be exploratory thinking that doesn't need systematic support yet. It might be a question the requester should answer themselves using self-service tools. Or it might be unclear thinking that needs refinement before consuming analytics capacity.

How This Changes the Conversation

Traditional request: "Can you pull customer satisfaction scores by region?"

With decision context required: Analyst: "What decision is this meant to inform?"

Requester: "We're trying to figure out where to focus service improvements."

Analyst: "Who's deciding on those improvements and when?"

Requester: "The VP of Operations. We have budget planning next month."

Analyst: "What are the options being considered?"

Requester: "Either we invest in the Southwest region where growth is highest, or we focus on the Northeast where scores are lowest."

Analyst: "So you need satisfaction scores by region, but specifically to compare Southwest performance against Northeast and recommend where service investment would have most impact?"

Requester: "Exactly. And it would help to know what's driving the low scores in Northeast so we can estimate improvement potential."

Now the analyst can build something focused: Regional satisfaction scores with Northeast root cause analysis and Southwest baseline, plus cost-impact scenarios for service improvements in each region. This directly enables the VP's decision next month about where to invest.

Same underlying question. Dramatically different and more useful output. All because decision context was established upfront.

Implementing Decision-Required Intake

Here's how to transform your intake process step by step.

Step 1: Update Your Intake Form

If you have a formal ticketing system or request form, add required fields:

Decision Context (Required)

- Decision to be made: [text field]
- Decision owner: [dropdown or text]
- Decision deadline: [date field]
- Options under consideration: [text field]
- Impact if decided poorly: [text field]

Work Type (Required)

- Decision support (supports specific decision from inventory)
- Exploratory analysis (hypothesis development, no immediate decision)
- Infrastructure work (data quality, system maintenance)
- Self-service enablement (building tools for others to use)

Make decision context required for "decision support" requests. For other work types, different information is needed, but make the classification explicit.

If you don't have a formal system, create a simple template that analysts send when receiving requests:

"Thanks for reaching out. To ensure I build something useful, help me understand:

- What decision is this meant to inform?
- Who's making that decision and when?
- What options are being considered?
- What would change based on what the analysis shows?"

Either way, establish that decision context comes before work begins.

Step 2: Train Requesters

Most people won't naturally think in terms of decisions and options. They think in terms of questions and curiosity. You need to teach them the new language.

Create a short guide (one page) that explains:

"Before requesting analytics work, please clarify the decision you're trying to make. This helps us build exactly what you need and ensures our limited capacity focuses on the highest-impact work.

Good request: 'I need to decide whether to invest $2M in customer retention or redirect to acquisition. Who owns this: VP of Customer Success. When: Decision needed by end of Q3 for Q4 planning. Options: Invest in retention program for top 50 at-risk accounts, or maintain current approach and fund acquisition instead. What I need: Comparison of retention program costs and expected results versus acquisition alternatives.'

Vague request: 'Can you give me some customer data?'

If you're not sure what decision you're supporting, that's okay! Let's talk through it together."

Run brief training sessions with frequent requesters. Walk through examples. Show how decision-framed requests get faster, better results.

Step 3: Establish Intake Triage

Assign one person (rotating weekly or monthly) to triage incoming requests. This person's job is to:

Review each request for decision context. Does it clearly specify the decision, owner, timing, and options? If yes, route to prioritization. If no, send back with the template questions.

Help clarify unclear requests. Schedule 15-minute conversations with requesters who are struggling. Walk them through the decision context questions. Often, this conversation reveals what's really needed.

Classify work type. Is this truly decision support, or is it exploratory, infrastructure, or better served by self-service? Route accordingly.

Reject or redirect requests that don't meet criteria. Some requests shouldn't consume analytics capacity at all. Self-service tools can answer them. The requester can get what they need from existing reports. Or the question isn't important enough to justify the effort.

Triage shouldn't be bureaucratic gatekeeping. It's quality control that protects both requesters and analysts from wasted effort.

Step 4: Implement Decision-Based Prioritization

With decision context captured, you can now prioritize rationally. Score requests on:

Strategic importance (1-5 scale) Does this decision directly influence strategic goals from your organization's priorities? Decisions in your Chapter 3 inventory automatically score high. Decisions that don't connect to strategy score low.

Decision consequence (1-5 scale) What's at stake? Large budget allocations, major risks, significant operational impacts score high. Small adjustments, routine operations score lower.

Decision frequency (1-5 scale) How often is this decision made? Weekly or monthly decisions compound their impact over time and score high. One-time decisions might be important but score lower on frequency.

Current quality gap (1-5 scale) How much better could this decision be with good analytics support? Decisions made with high confidence using good information don't need more support. Decisions made with frequent delays, low confidence, or poor outcomes score high.

Timing urgency (1-5 scale) When must the decision be made? Next week scores higher than next quarter, but only if the decision consequence also justifies urgency.

Multiply or sum these scores to get overall priority. Requests scoring in the top tier get immediate attention. Middle tier gets scheduled. Bottom tier gets deferred or redirected to self-service.

The specific scoring method matters less than having explicit, consistent criteria that everyone understands.

Step 5: Communicate the Change

Send clear communication to stakeholders about the new process:

"**Changes to Analytics Request Process**

To ensure our analytics capacity focuses on the highest-impact work and to deliver better results faster, we're implementing decision-focused intake.

What's changing: All analytics requests will now require clear decision context: what decision you're making, who owns it, when it's needed, and what options you're considering.

Why this matters: This lets us build exactly what you need, prioritize work based on strategic impact, and deliver focused analysis instead of comprehensive reports that miss the point.

What you should do: Before requesting analytics work, clarify your decision. Use our new request template [link]. If you're not sure how to frame it, schedule 15 minutes with [triage person] and we'll help.

What you'll get: Faster turnaround on high-priority decisions, better-focused analysis, and confidence that our team is working on what matters most to the organization."

Expect questions and some resistance. Address it directly in office hours, team meetings, and one-on-one conversations.

Handling Common Objections

You'll face predictable pushback. Here's how to respond.

"Not every request is about a decision"

Response: "You're right. Some work is exploratory, some is infrastructure, some is self-service enablement. Our new process classifies these different work types so each gets appropriate treatment. But if you're asking for decision support, we need to understand the decision."

"This will slow down requests"

Response: "It might add 15 minutes upfront. But it eliminates days of rework when we build the wrong thing because we didn't understand what you really needed. And it helps us prioritize so truly urgent decisions get immediate attention instead of waiting behind less important work."

"I don't always know what decision I'm supporting yet"

Response: "That's completely fine. Let's talk it through. Often, that conversation clarifies what you actually need and whether analytics is the right solution. If you're just exploring, we can point you to self-service tools or suggest a quick scoping analysis before committing to a major project."

"My executive just wants this, I can't push back"

Response: "We're not asking you to push back. We're offering to help you clarify what your executive really needs. Often when an executive asks for 'customer data,' they're trying to make a specific decision but haven't articulated it yet. We can help you figure out what they're really asking for, which makes them happier with the result."

"This seems like bureaucracy"

Response: "It's quality control. Just like we wouldn't start building software without requirements, we shouldn't start analysis without understanding the decision. It's actually less bureaucratic than endless revision cycles when we build the wrong thing."

The key is positioning this as serving requesters better, not creating obstacles.

Before and After Examples

Here's how the same requests look before and after implementing decision-required intake:

Example 1: Workforce Request

Before: Request: "Can you give me turnover data?" Analyst builds: Comprehensive turnover report by department, role, tenure, and demographics Result: Requester glances at it, doesn't use it, and later asks for different cuts of the data

After: Request: "I need to decide where to allocate hiring budget across three regions. Who: Regional Operations VP. When: Next month for Q4 planning. Options: Invest heavily in Northeast where growth is highest, spread evenly across regions, or focus on Southeast where turnover is most problematic. I need: Turnover by region, time-to-fill by region, cost of vacancy by region, and impact analysis of different budget allocation scenarios." Analyst builds: Focused analysis showing turnover, hiring difficulty, and vacancy costs by region, with three budget scenarios and expected outcomes Result: VP uses this in planning meeting, makes clear decision on Northeast focus

Example 2: Project Portfolio Request

Before: Request: "We need a dashboard for project portfolio management" Analyst builds: Comprehensive dashboard with dozens of metrics, beautiful visualizations, extensive filtering Result: Used occasionally for general awareness but doesn't change how portfolio decisions are made

After: Request: "We make three types of portfolio decisions: quarterly funding allocation, monthly resource reallocation, and exception escalation when projects hit trouble. Who: Portfolio Director for funding, PM leads for resources, PMO Director for escalations. When: Quarterly, monthly, and triggered by variance thresholds. Options vary by decision type but all need to compare current state against targets and available alternatives. I need: Three focused views, one per decision type, showing only the metrics that inform each specific choice." Analyst builds: Three decision-specific views with clear thresholds that trigger actions Result: All three decision types now use the dashboard systematically, portfolio performance improves

Example 3: Customer Analysis Request

Before: Request: "Can you analyze our customer satisfaction trends?" Analyst builds: Trend analysis with segmentation, correlation analysis, benchmark comparisons Result: Interesting findings presented in meeting, generates discussion, no clear next steps

After: Request: "We need to decide where to invest $500K in service improvements. Who: VP of Operations. When: Decision in two weeks for annual budgeting. Options: Invest in phone support capacity, invest in self-service tools, invest in field service response time, or split the investment. I need: Current satisfaction by channel, cost to improve each channel, expected satisfaction impact of improvements, and customer lifetime value by satisfaction level to estimate ROI." Analyst builds: Channel satisfaction analysis with cost-benefit scenarios for each investment option Result: VP chooses based on ROI analysis, investment decision made in budgeting meeting

The pattern is consistent. Decision context transforms vague requests into focused work that actually influences choices.

Measuring Success

Track these metrics to know if your intake transformation is working:

Percentage of requests with clear decision context Start: Probably below 20% Target: Above 80% within three months

Average time from request to delivery for high-priority work Should decrease as you eliminate low-priority work and reduce rework cycles

Rework rate (requests requiring significant revision after initial delivery) Start: Probably 30-50% Target: Below 10%

Percentage of analytics work that influences documented decisions Start: Difficult to measure but likely below 40% Target: Above 70%

Requester satisfaction with delivered work Should increase as you deliver what's actually needed

Analyst satisfaction with work meaningfulness Should increase as more work influences real decisions

Survey requesters quarterly: "Did the analysis we provided help you make a better decision?" Track the yes rate.

Survey analysts quarterly: "What percentage of your work this quarter influenced important decisions?" Track the trend.

Integration With Your Strategy Layer

Decision-required intake becomes powerful when it connects to the Strategy Layer components you've built.

Linking to Your Decision Inventory

When requests arrive, the first question is: "Is this decision in our inventory?"

If yes, you already have documentation about who owns it, how often it occurs, what metrics should inform it, and what strategic goal it serves. The request gets prioritized based on the decision's importance.

If no, you ask: "Should this decision be in our inventory?" Maybe it's a new strategic priority that wasn't previously cataloged. Add it. Or maybe it's not actually strategic enough to warrant inventory inclusion, which helps calibrate prioritization.

The inventory becomes your reference for what deserves systematic support versus what's nice to have.

Using Your Decision-Critical Metrics

When requests specify decisions from your inventory, you can often respond: "We already have metrics for that decision. Here's what they currently show. Do you need additional analysis, or does this answer your question?"

This prevents duplicate work and ensures consistency. Multiple people asking about the same decision all get the same underlying metrics, just presented for their specific use case.

Creating Systematic Learning

When you consistently document what decision was being supported, what analysis was provided, and whether it influenced the choice, you build institutional knowledge.

Over time, you learn which types of analysis are most influential for which types of decisions. You develop patterns you can reuse. You identify decisions that consistently need more support and can proactively build better infrastructure for them.

This learning compounds. Your team gets better at decision support because you're tracking what works.

Your Request Intake Template

Here's a template you can adapt for your organization:

ANALYTICS REQUEST FORM

Requester Information

- Name:
- Department:
- Email:
- Date submitted:

Work Type (check one) ☐ Decision support (inform specific decision) ☐ Exploratory analysis (hypothesis development) ☐ Infrastructure (data quality, systems) ☐ Self-service enablement (building tools)

For Decision Support Requests (required):

Decision Context

- Decision to be made:
- Decision owner (specific person or role):
- Decision deadline:
- Options being considered:
- What changes based on analysis results:

Current Situation

- What prompted this request:
- What you already know:
- What's unclear or unknown:

Desired Outcome

- What specific questions need answering:
- How results should be presented:
- Who else needs to see this:

Strategic Connection

- Which strategic goal does this decision influence:
- Estimated impact if decided well:
- Consequence if decided poorly:

For Other Work Types: [Different fields appropriate to exploratory, infrastructure, or enablement work]

Priority Self-Assessment (we'll validate) ☐ High (decision in next 2 weeks with major impact) ☐ Medium (decision in next month with moderate impact) ☐ Low (decision timing flexible or impact limited)

Implementation Exercise: Redesign Your Intake Process

Time required: 2-3 hours

What you'll create: Updated intake process with decision-required fields

Steps:

1. Document current state (30 minutes)

How do requests arrive today? Email, Slack, ticketing system, hallway conversations? What information is captured? What's missing?

Create a simple flow diagram showing how work enters your team now.

2. Design new intake form (45 minutes)

Using the template above as a starting point, create your version. Include:

- Decision context fields (required for decision support work)
- Work type classification
- Strategic connection questions
- Priority assessment guidance

Keep it simple enough that people will actually use it. One page maximum.

3. Identify triage approach (15 minutes)

Who will review incoming requests for completeness? How often will they do this? What authority do they have to send requests back or redirect them?

Will this be a rotating role among senior analysts, or a dedicated function?

4. Draft communication plan (30 minutes)

Write the message you'll send to stakeholders announcing the change. Include:

- What's changing and why
- What requesters need to do differently
- What they'll get in return (better, faster results)
- Who to contact with questions

- Timeline for rollout

Also plan how you'll train frequent requesters. Office hours? Workshops? One-on-one sessions?

5. Pilot with friendly users (varies)

Before full rollout, test with 3-5 requesters who are likely to be supportive. Have them use the new process for their next request. Learn what's confusing, what needs adjustment, what works well.

Refine based on feedback.

What you have now: A complete intake redesign ready to implement. You've thought through the mechanics, the communication, and the transition approach.

Next step: Implement with your team. Start using the new process for all incoming requests. Expect a learning curve of 2-4 weeks as people adjust. Celebrate early wins where decision context led to better results.

Moving Forward

You've now built the core operational components of your Strategy Layer:

Chapter 3: Decision inventory (cataloged important choices) **Chapter 4:** Decision-critical metrics (measurements that inform choices)**Chapter 5:** Decision briefs (format for presenting to executives) **Chapter 6:** Decision-required intake (ensuring analytics work connects to decisions)

These pieces create a complete system for connecting analytics work to strategic decisions. Work flows to what matters. Metrics serve clear purposes. Executives get what they need to choose. Decisions improve systematically.

The next chapters show you how to apply this system to organizational challenges where it creates breakthrough clarity: cross-functional alignment where metrics conflict, AI deployment where decision context is essential, and enterprise scaling where consistent practices create competitive advantage.

Chapter 7 tackles one of the hardest problems in any organization: getting different functions to align around shared decisions when their metrics point in different directions.

The Strategy Layer becomes the shared language that makes this possible.

You're ready to solve cross-functional conflicts.

Chapter 7: Cross-Functional Alignment

A manufacturing company wanted to improve equipment reliability. Operations defined reliability as mean time between failures. Maintenance defined it as preventive maintenance completion rates. Finance defined it as unplanned maintenance cost. Quality defined it as defect rates attributable to equipment issues.

Each function built metrics, held teams accountable to different targets, and believed the others weren't taking reliability seriously enough. When the VP of Manufacturing asked why reliability wasn't improving despite everyone's focus on it, the answer was simple: the organization was optimizing four different things and calling them all "reliability."

No amount of effort or investment could improve an outcome no one had defined consistently.

This pattern repeats constantly. Strategic goals that require cross-functional execution fragment into function-specific interpretations, metrics, and priorities. Each function works hard within its domain, but the organization as a whole doesn't improve because the functions aren't aligned toward the same outcome.

This chapter shows how the Strategy Layer becomes a shared language across functions, enabling genuine alignment rather than just coordination or compromise.

The Cross-Functional Conflict Pattern

Most strategic outcomes depend on decisions and actions from multiple functions. Customer experience depends on sales, service, product, and operations. Financial performance depends on every function's cost management and revenue contribution. Quality depends on design, manufacturing, supplier management, and inspection.

When functions aren't aligned, predictable conflicts emerge.

Competing KPIs and Local Optimization

Each function defines success in terms of their own metrics, which often conflict with other functions' metrics.

Sales measures revenue and new customer acquisition, which can encourage bringing in customers who aren't good fits. Service measures resolution time and customer satisfaction, which suffers when sales brings in poorly matched customers. Operations measures throughput and efficiency, which conflicts with customization requests from sales.

Each function hits their targets while organizational performance stagnates because local optimization doesn't create system-wide optimization.

Sequential Blame and Handoff Dysfunction

When outcomes are poor, functions blame whoever came before them in the process.

Sales blames marketing for weak leads. Operations blames sales for unrealistic commitments. Service blames operations for quality issues. Finance blames everyone for cost overruns.

These blame chains distract from the real issue: the functions aren't working as a system toward shared outcomes. Handoffs between functions become adversarial rather than collaborative.

Invisible Trade-offs Made Unilaterally

Every function makes decisions that involve trade-offs affecting other functions, but these happen invisibly.

IT decides to defer infrastructure investment, creating technical debt that constrains future product development. HR decides to centralize recruiting, slowing hiring for critical operational roles. Procurement decides to consolidate vendors, reducing flexibility for specialized needs.

These decisions may be rational within each function's context but impose costs elsewhere that weren't explicitly considered.

Dueling Narratives in Executive Meetings

Cross-functional conflicts surface in leadership meetings as competing interpretations of the same situation.

Operations says manufacturing performance is strong based on throughput metrics. Quality says it's declining based on defect rates. Both are technically correct using their chosen metrics, but they can't both be right about whether things are getting better or worse.

Executives waste time arbitrating these disputes rather than focusing on what to do differently.

Resource Competition Without Shared Criteria

When budget, headcount, or leadership attention is constrained, functions compete for resources. Without shared frameworks for evaluating trade-offs, these competitions get resolved through politics, negotiating skill, or whoever presented most recently.

Strategic priorities should determine resource allocation, but without shared language and metrics, strategic priority becomes whatever argument each function makes.

The Strategy Layer as Shared Language

The Strategy Layer addresses cross-functional conflict by creating one shared definition of success while acknowledging that different functions contribute through different means.

The key insight: there should be **one Strategy Layer per strategic outcome**, with different functions plugging in their metrics and constraints, but all aligned on the same decisions.

Without a Strategy Layer: Fragmented Approach

Consider customer retention as a strategic outcome.

Sales tracks new bookings. Marketing tracks leads and campaign engagement. Service tracks resolution time and satisfaction scores. Product tracks feature adoption. Finance tracks revenue and cost to serve.

Each produces their own reports and dashboards. When retention declines, each function has a theory about causes (all focused on other functions' failings) and proposed solutions (all requiring other functions to change while their own function continues current approach).

With a Strategy Layer: Aligned Approach

The organization identifies the key decisions that drive retention: which at-risk customers to target for intervention, what retention tactics to deploy for different customer types, how to allocate retention investment across segments, when to accept churn versus fight for renewal.

For each decision, they define the questions that need answers. Then each function contributes the metrics and insights from their domain that inform these shared decisions.

Sales provides data on customer expansion patterns. Marketing provides campaign response rates. Service provides satisfaction and issue trends. Product provides usage and feature adoption. Finance provides profitability and lifetime value.

The difference is profound. In the first scenario, each function optimizes independently and coordination happens through meetings, negotiations, and escalations. In the second scenario, functions work toward explicit shared decisions using a common framework, and their different metrics complement rather than compete.

Building Shared Decision Frameworks

Here's how to create alignment through the Strategy Layer.

Step 1: Convene Stakeholders Around One Outcome

Pick one strategic outcome that requires cross-functional execution. Not "improve everything" but something specific: customer retention, operational efficiency, product quality, time to market, employee engagement.

Bring together representatives from each function that significantly influences this outcome. The goal isn't consensus on every detail but alignment on the decision framework.

Example: Hospital Patient Throughput

Strategic outcome: Move patients efficiently through care while maintaining quality

Functions involved: Emergency department, inpatient units, surgical services, ancillary services, nursing, care management, capacity planning

Stakeholders: Representatives from each area, plus someone who can facilitate the Strategy Layer design (often from strategy, operations, or analytics)

Step 2: Map Shared Decisions

Together, identify the critical decisions that drive the outcome. Not every decision related to the outcome, just the repeating, high-impact choices where alignment would create the most value.

For patient throughput:

- How to allocate beds across services each day
- When to implement surge protocols
- How to prioritize surgical scheduling
- Where to invest in capacity improvements
- How to staff units based on predicted census and acuity

These decisions span multiple functions. Emergency department can't optimize throughput alone. Inpatient units need coordination with surgical services. Capacity planning depends on nursing availability.

Making these decisions explicit shows where alignment is necessary.

Step 3: Define Questions for Each Decision

For each shared decision, identify what needs to be known to decide well. This is where different functional perspectives combine.

Decision: How to allocate beds across services each day

Questions from different functions:

- What's current and anticipated census by service? (Capacity planning)
- Which patients are ready for discharge? (Care management)
- What surgical cases are scheduled? (Surgical services)

- What staffing is available by unit? (Nursing)
- What's the ED pipeline? (Emergency department)
- What's our bed utilization target? (Finance)

Each function contributes questions from their domain. Together, these questions cover what's needed to make the allocation decision well.

Step 4: Identify Metrics That Answer Questions

Now different functions propose metrics that answer the shared questions.

For bed allocation decision:

From capacity planning: Current occupied beds by service, available beds by unit, historical utilization patterns

From care management: Discharge readiness status, average length of stay by diagnosis

From surgical services: Scheduled procedures by service, estimated procedure duration, post-op bed requirements

From nursing: Staffing levels by unit, nurse-to-patient ratios, overtime usage

From emergency: Current ED census, expected admissions next 8/16/24 hours, acuity distribution

From finance: Cost per bed day, revenue by service line, contribution margin by patient type

Each function provides metrics from their expertise. But they're all organized around the shared decision about daily bed allocation, not optimizing for separate functional goals.

Step 5: Establish Shared Thresholds and Trade-off Rules

With metrics identified, define together what different levels mean and how to resolve conflicts when metrics point in different directions.

Threshold example:

- When overall occupancy exceeds 95%, activate surge protocol
- When any service exceeds 100% capacity, prioritize that service for next available bed regardless of other factors
- When nurse staffing falls below 1:5 ratio for high-acuity patients, defer elective procedures until staffing improves

Trade-off rule example: "When deciding between service lines competing for limited beds, prioritize in this order: (1) patient safety (never compromise care), (2) emergency admissions (legally and ethically required), (3) scheduled surgical cases (high revenue, contractual obligations), (4) elective admissions (flexible timing)."

These rules make trade-offs explicit rather than leaving them to be negotiated case by case or decided by whoever has the most organizational power.

Step 6: Create Shared Review Forum

Establish a regular meeting where stakeholders examine performance across the shared Strategy Layer.

Not separate meetings where each function presents their metrics. A joint review where shared decisions are examined: Are we making these decisions on schedule? Do we have the metrics we need? Are outcomes improving? What should we adjust?

For patient throughput:

Weekly operations meeting with representatives from all relevant functions. Agenda:

- Review past week's bed allocation decisions and outcomes
- Examine throughput metrics (length of stay, ED wait times, bed utilization)
- Identify where the system isn't working (bottlenecks, coordination failures)
- Adjust allocation rules or capacity plans as needed
- Ensure all functions have the information they need for next week's decisions

This forum reinforces that functions share accountability for outcomes, not just for their departmental metrics.

Examples Across Decision Types

Cross-functional alignment through the Strategy Layer looks different in different contexts, but the pattern remains consistent.

IT and Finance: Project Portfolio Decisions

Typical conflict: IT wants to invest in platforms and capabilities. Finance wants to minimize spending and maximize short-term returns. Portfolio reviews become negotiations where each side makes their case.

Strategy Layer approach:

Shared decisions: Which projects get funded this quarter? Which get delayed? Which get stopped?

Shared questions: What's the business value? What are technical dependencies? What's at risk if unfunded? What capability are we building?

Metrics from IT: Technical debt scores, platform risk levels, dependency mapping, capability gaps

Metrics from Finance: Cost, ROI projections, budget variance, cash flow impact

Metrics from business units: Value delivery scores, urgency rankings, strategic alignment

Trade-off rules: "Technical debt reduction gets 20% of budget floor. Revenue-generating projects get priority in remaining budget unless technical debt threatens platform stability. Projects without clear business sponsor get deferred regardless of technical merit."

Result: Portfolio reviews where IT and Finance use the same framework. They can disagree about conclusions (which is healthy) without disagreeing about facts or priorities. The CFO understands why certain technical projects matter strategically. The CIO understands financial constraints and can articulate technology decisions in business terms.

Operations and HR: Staffing Decisions

Typical conflict: Operations needs adequate staffing to meet service levels and quality targets. HR needs to manage labor costs and avoid churning through staff with unsustainable workload expectations.

Strategy Layer approach:

Shared decisions: Where to allocate hiring budget across locations and roles? When to hire versus using overtime or contractors? How to prioritize recruiting effort? What compensation to offer for hard-to-fill positions?

Shared questions: Where are we chronically understaffed relative to workload? Where is workload declining and staffing could be reduced? Where is turnover highest and why? What are market rates for critical skills? What's the cost of understaffing versus overstaffing?

Metrics from Operations: Workload by location, throughput rates, overtime usage, service level achievement, quality incidents attributed to staffing

Metrics from HR: Turnover by role and location, time-to-fill, compensation competitiveness, employee engagement scores, exit interview themes

Trade-off rules: "Roles directly impacting customer experience or safety get priority for hiring even if above budget. Administrative roles defer hiring until operational roles are adequately staffed. Compensation can exceed market 75th percentile for roles with demonstrable business impact and recruiting difficulty."

Result: Staffing decisions serve both operational needs and workforce sustainability. Operations gets adequate staffing for strategic priorities. HR maintains manageable costs and healthy work environments. Trade-offs are made explicitly based on shared understanding rather than implicitly through organizational friction.

Clinical and Financial: Care Delivery Decisions

Typical conflict: Healthcare organizations face constant tension between clinical quality and financial sustainability. Clinicians prioritize patient outcomes. Finance worries about whether the organization can continue operating if costs consistently exceed revenue.

Strategy Layer approach for cardiovascular service line:

Shared decisions: Which patients to accept for procedures (matching clinical need with appropriate setting)? How to configure staff and resources across sites? What quality improvement initiatives to fund? How to manage length of stay and readmissions?

Shared questions: What's the appropriate clinical setting for different patient types? What staffing models optimize both outcomes and efficiency? Which quality improvements have highest impact relative to cost? What drives length of stay and readmissions?

Metrics from Clinical: Outcomes by procedure type, complications, readmissions, patient mix complexity, clinical protocols adherence, quality scores

Metrics from Finance: Cost per case by procedure, margin by procedure type, resource utilization, revenue cycle performance, payer mix

Trade-off rules: "Patient safety overrides financial considerations in all cases. For elective procedures, cases are accepted when clinical appropriateness and financial viability both meet threshold criteria. Quality improvement investments require demonstrated clinical benefit AND financial sustainability within 24 months."

Result: Shared framework makes difficult conversations productive. Everyone works from shared data toward shared decisions. Clinical and financial perspectives both inform choices rather than one dominating or the two remaining in conflict.

Product, Engineering, and Security: Feature Development

Typical conflict: Product wants features that delight customers and drive revenue. Engineering wants time to address technical debt and improve platform stability. Security wants controls and compliance that protect the organization.

Strategy Layer approach:

Shared decisions: What features to prioritize in next release? What technical debt to address? Which security controls to implement? How to balance new capability versus stability versus protection?

Shared questions: What customer needs are most urgent? What technical debt is creating the most friction? What security vulnerabilities pose the highest risk? What's the opportunity cost of delaying features versus the risk of accumulating technical debt?

Metrics from Product: Customer feature requests by volume and revenue impact, competitive gaps, user engagement with existing features, retention correlation with feature sets

Metrics from Engineering: Technical debt severity scores, system stability metrics, development velocity, platform scalability headroom

Metrics from Security: Vulnerability counts by severity, compliance status, incident frequency, control coverage gaps

Trade-off rules: "Critical security vulnerabilities must be addressed before new feature work. Technical debt gets minimum 20% of development capacity. Feature prioritization balances customer revenue impact (50%), competitive necessity (30%), and technical enablement of future work (20%)."

Result: Development roadmaps that serve business needs while maintaining platform health and security. Product gets features that matter most. Engineering gets time to address technical sustainability. Security gets appropriate protection. Conflicts still occur, but they're resolved through explicit frameworks rather than political battles.

Alignment Mechanisms That Sustain

Creating initial alignment is one thing. Maintaining it as strategy and circumstances evolve requires deliberate mechanisms.

Joint Decision Workshops

When starting to build shared Strategy Layers, bring stakeholders together for focused design sessions. Half-day workshops where you collectively work through decisions, questions, metrics, and trade-offs.

These workshops build shared understanding and create co-ownership. People are more committed to frameworks they helped design than frameworks imposed on them.

Workshop structure:

- Hour 1: Map shared decisions for the strategic outcome
- Hour 2: Define questions each decision requires
- Hour 3: Identify metrics from each function that answer questions
- Hour 4: Establish thresholds and trade-off rules

Document the output. This becomes your shared Strategy Layer for that outcome.

Pre-Agreed Trade-off Rules

Some conflicts are predictable. Rather than negotiating them each time they arise, establish standing rules.

Common trade-off patterns:

Cost versus quality: "We optimize for quality up to $X cost per unit, beyond which cost concerns take precedence unless exception is approved by [role]."

Speed versus risk: "We move fast on reversible decisions, slow on irreversible decisions. [Definition of each category provided]."

Customer versus employee needs: "We prioritize employee safety over customer convenience, but optimize for both wherever possible."

Short-term versus long-term: "We accept up to Y% short-term margin pressure for initiatives with credible Z-month payback."

Document these rules. Reference them when conflicts arise. Update them as strategy evolves or as you learn from experience.

Escalation Paths When Metrics Clash

Even with good frameworks, situations arise where different functions' metrics lead to different conclusions about what to do. Define how to resolve these conflicts.

Clear escalation path components:

Threshold for escalation: What level of conflict triggers escalation versus being resolved among stakeholders directly? "When functions can't reach agreement within one business day" or "When decision involves more than $X in investment or risk."

Escalation owner: Who arbitrates escalated decisions? Often the senior leader accountable for the strategic outcome both functions serve.

Required information: What must be presented? Both perspectives, metrics supporting each position, trade-off implications, recommendations from each side.

Timeline: How quickly are escalated decisions resolved? "Within 48 hours" prevents conflicts from lingering and blocking work.

Example from financial services:

"When Operations and Risk disagree about a process change, they first attempt to find solutions that serve both functions. If no solution emerges within one week, the issue escalates to the COO with a structured brief showing Operations' efficiency metrics, Risk's control metrics, and recommendations from each side. The COO decides within 48 hours based on strategic priorities and risk appetite."

This structure turns conflicts that previously consumed weeks into decisions made in days.

Shared Review Forums

Regular joint meetings where stakeholders examine performance across the shared Strategy Layer create ongoing accountability.

Effective forum characteristics:

Cross-functional attendance: Representatives from all functions that influence the outcome, not just the one that called the meeting.

Decision focus: Agenda organized around the shared decisions in your Strategy Layer, not functional reporting.

Outcome orientation: Discussion centers on whether the strategic outcome is improving and what's blocking progress, not on defending functional metrics.

Forward looking: More time on what to adjust going forward than on explaining past performance.

Action orientation: Every meeting ends with clear commitments about what will change.

Example agenda for customer retention forum:

1. Review retention metrics for past quarter (10 minutes)
2. Examine each key decision: Are we making them on schedule? Do we have needed information? (20 minutes)
3. Identify disconnects: Where are functions out of sync? Where are metrics sending conflicting signals? (15 minutes)
4. Decide on adjustments: What should change in how we make these decisions? (10 minutes)
5. Commitments: Who will do what by when? (5 minutes)

Total: 60 minutes, once per quarter, focused on shared decisions rather than functional reporting.

Making Alignment Visible

Alignment works best when it's visible to the organization, not just understood by the people in the room.

Shared Dashboards

Create dashboards that show the metrics for shared decisions, organized by decision rather than by function.

Not separate dashboards where Sales shows sales metrics, Service shows service metrics, and Product shows product metrics. One shared dashboard that shows all the metrics relevant to customer retention decisions, regardless of which function owns which metric.

This visibility prevents functions from optimizing for their own metrics while ignoring the shared outcome.

Shared Documentation

Document your Strategy Layer for each cross-functional strategic outcome. Make this documentation accessible to anyone who needs to understand how decisions are made.

Include:

- The strategic outcome and why it matters
- The critical decisions that drive the outcome
- Which functions contribute to each decision
- The metrics each function provides
- The trade-off rules that guide resolution of conflicts
- Who owns overall coordination

This documentation serves multiple purposes. New employees understand how the organization approaches this outcome. People can see how their work fits into the larger system. When conflicts arise, there's a shared reference point.

Celebration of Shared Success

When alignment creates visible results, celebrate it explicitly as cross-functional success, not individual functional achievement.

"Customer retention improved 12% this year because Sales targeted the right accounts, Service addressed the right issues, Product built the right features, and Finance allocated resources appropriately. This is what alignment looks like."

This recognition reinforces that shared outcomes require shared effort and creates motivation for continued collaboration.

Knowing When Alignment Is Working

Several signals indicate whether your cross-functional Strategy Layer is creating genuine alignment:

Functions reference the same metrics when discussing the outcome. When retention comes up in any meeting, everyone talks about the same churn risk scores, retention program success rates, and customer lifetime value calculations, not competing sets of metrics.

Trade-offs get resolved through frameworks, not escalation. When cost and quality conflict, people apply the established trade-off rules rather than escalating to senior leadership or engaging in political battles.

Decisions happen faster. Cross-functional decisions that previously took weeks of negotiation now take days because the framework makes it clear how to evaluate options and resolve disagreements.

Blame patterns decrease. Functions spend less time pointing fingers at each other and more time problem-solving together because they share accountability for outcomes.

New people can understand quickly. When someone joins the organization or moves into a new role, they can review the shared Strategy Layer and understand how decisions are made in this domain without needing to learn through experience or organizational politics.

Strategic outcomes improve. The ultimate test is whether the alignment produces better results. Is retention improving? Is quality increasing while costs stay reasonable? Are projects delivering faster with maintained standards?

If you're not seeing these signals, investigate where the Strategy Layer implementation is incomplete or where old patterns are persisting despite the new framework.

Implementation Exercise: Build One Shared Strategy Layer

Time required: Half-day workshop plus 2 hours follow-up

What you'll create: Complete shared Strategy Layer for one cross-functional outcome

Steps:

1. Select the outcome and stakeholders (before workshop)

Choose one strategic outcome that requires cross-functional execution where alignment is currently weak. Identify which functions significantly influence this outcome. Invite representatives from each.

2. Workshop Session 1: Map decisions (60 minutes)

Together, list the critical decisions that drive the outcome. Use sticky notes or whiteboard. Each person contributes from their functional perspective.

Cluster related decisions. Prioritize to the top 3-5 that matter most.

For each priority decision, document: What's being decided? How often? Who should own this?

3. Workshop Session 2: Define questions (60 minutes)

For each priority decision, identify the questions that need answering. Go around the room, each function contributing questions from their perspective.

Write all questions on the board. Cluster similar questions. Prioritize to the essential 3-5 questions per decision.

4. Workshop Session 3: Identify metrics (60 minutes)

For each question, identify which metrics answer it. Have each function propose metrics they can provide from their domain.

Document metric definitions, data sources, and which function owns each metric.

Identify gaps: questions that don't have good metrics yet.

5. Workshop Session 4: Establish rules (60 minutes)

Define thresholds for key metrics: what values trigger what actions?

Establish trade-off rules for predictable conflicts: when metrics point in different directions, how do we resolve it?

Specify escalation path: when functions can't resolve disagreements, who decides and how?

6. Post-workshop: Document and implement (2 hours)

Create a clear document capturing:

- Strategic outcome
- Critical decisions
- Questions per decision
- Metrics per question with ownership
- Thresholds and trade-off rules
- Review forum structure

Share with all participants. Schedule first shared review forum.

What you have now: A functioning shared Strategy Layer for one cross-functional outcome. Functions now have shared language, shared metrics, and shared frameworks for resolving conflicts.

Next steps:

- Run the shared review forum quarterly
- Track whether alignment improves (faster decisions, less conflict, better

outcomes)

- Extend the approach to other cross-functional outcomes based on lessons learned

Moving Forward

Cross-functional alignment is one of the hardest challenges in organizational life. Different functions have different expertise, different perspectives, and different metrics. These differences are valuable. The challenge is channeling them toward shared outcomes rather than letting them fragment into competing priorities.

The Strategy Layer provides the structure that makes this possible. Not by forcing everyone to care about the same things equally, but by creating shared language around the decisions that require coordination.

You've now built the complete operational framework:

Chapters 3-4: Decision inventory and metrics **Chapters 5-6:** Decision briefs and analytics intake **Chapter 7:** Cross-functional alignment through shared Strategy Layers

The next chapters show you how to scale this approach: deploying AI within your Strategy Layer frameworks, developing leaders who can champion decision-centric approaches, and building enterprise capability that creates sustained competitive advantage.

Chapter 8 addresses a question on every organization's mind: where does AI fit in all this?

The answer: AI accelerates good decision-making when deployed within clear decision frameworks. Without those frameworks, it just automates confusion faster.

You're ready to learn how to use AI strategically.

Chapter 8: AI Within the Strategy Layer

Artificial intelligence dominates technology conversations today. Organizations launch AI initiatives, hire data scientists, experiment with machine learning models, and deploy generative AI tools. Vendors promise that AI will revolutionize decision making, automate complex analysis, and uncover insights humans would miss.

Some of these promises are real. AI capabilities have advanced dramatically. But AI doesn't eliminate the need for the Strategy Layer. It makes the Strategy Layer more important.

This chapter clarifies how AI fits within your decision frameworks. The goal is helping you deploy AI where it creates genuine value while avoiding the common trap of automating confusion or generating impressive insights that no one acts on.

Where AI Actually Helps

Understanding AI's role starts with recognizing what current AI technologies do well, stripped of marketing claims.

Pattern Detection in Complex Data

AI excels at finding patterns across more variables, more complex relationships, and larger datasets than humans can process.

Machine learning models can identify which combinations of customer behaviors predict churn, which project characteristics forecast delays, which patient factors indicate readmission risk, or which operational conditions lead to quality defects. These patterns may be subtle, nonlinear, or interactive in ways that traditional methods miss.

This capability is valuable when you have enough data to train models, when patterns actually exist (not all situations are predictable), and when detecting patterns earlier or more accurately would change decisions.

Example: An AI model that predicts equipment failures three days earlier than current methods enables better maintenance scheduling. An AI model that's marginally more accurate but delivers results too late to act isn't useful regardless of technical sophistication.

Forecasting and Scenario Analysis

AI models generate predictions about future states based on historical patterns and current conditions. Sales forecasts based on pipeline data, demand forecasts based on seasonal patterns, capacity forecasts based on workload trends, or financial forecasts based on operational drivers.

AI can also simulate scenarios showing what happens if you change pricing by 10%, if a competitor enters your market, if regulatory requirements shift, or if you reallocate resources from one area to another.

These capabilities matter when decisions depend on forward-looking views and when AI forecasts are meaningfully better than simpler alternatives. Many situations don't require AI-powered forecasts. Simple trend extrapolation or domain expert judgment works fine. AI adds value when relationships are complex enough that sophisticated modeling improves accuracy materially.

Anomaly Detection and Alerting

AI monitors complex systems and flags unusual patterns that might indicate problems. Transactions that look like fraud, system behaviors suggesting impending failures, operational metrics deviating from normal patterns in ways predicting quality issues, or employee behaviors correlating with attrition risk.

The advantage over rule-based alerting is that AI learns what "normal" looks like across many variables and detects subtle anomalies that fixed rules would miss.

This matters when anomalies are important to catch early, when they're subtle enough that rule-based approaches produce too many false positives, and when someone can actually investigate and respond to alerts. An AI system flagging 100 potential issues per day when analysts can investigate 5 creates alert fatigue, not value.

Optimization for Complex Trade-offs

AI searches vast solution spaces to find combinations that optimize multiple objectives with complex constraints. Optimal scheduling for staff or equipment, optimal routing for logistics, optimal product configurations for manufacturing, or optimal resource allocation across competing priorities.

These problems have too many possible solutions for humans to evaluate exhaustively, but AI can find high-quality solutions quickly.

This matters when the optimization problem is genuinely complex (not solvable with simpler heuristics), when the optimal solution creates meaningful value over good-enough solutions, and when constraints and objectives can be specified clearly. You can't optimize for vaguely defined goals.

What AI Cannot Replace

The limitations of AI aren't primarily technical. They're structural. AI can't replace the human work of defining what matters and what to do about it.

Clear Strategic Goals

AI can optimize, but it can't determine what to optimize for.

A machine learning model can predict customer churn with high accuracy, but it can't decide whether retention or acquisition should be the strategic priority, whether all churn is equally important to prevent, or what trade-offs between retention cost and customer lifetime value are acceptable.

These are strategic choices humans must make. Organizations that deploy AI without clear strategic direction end up optimizing for whatever is easy to measure rather than what actually matters.

An AI-powered recommendation engine might maximize click-through rates while degrading customer trust. An AI scheduling system might maximize utilization while burning out staff. The AI is working exactly as designed. The problem is the design reflects unclear strategy.

Well-Defined Decisions

AI can inform decisions, but it can't replace the work of defining what decisions exist, who owns them, and what constitutes success.

If your organization hasn't clarified its critical decisions (Chapter 3), AI will either be applied to wrong decisions or generate insights that don't map to any decisions at all.

A financial services firm deployed sophisticated AI models analyzing customer data, producing detailed risk scores and propensity models. The models were technically excellent. But the organization hadn't defined the decisions these models should support: which customers to contact, what offers to make, what interventions to trigger.

The models sat unused because no one had connected them to actual decision processes with clear ownership and defined actions.

Accountable Ownership

AI can provide recommendations, but someone must be accountable for decisions that follow those recommendations.

"The AI decided" is never an acceptable answer when something goes wrong. Humans own decisions, even when AI informs them. This ownership includes understanding what the AI is doing well enough to judge when to trust its recommendations and when to override them.

Organizations that blur this accountability create risk. When AI recommendations are followed automatically without human judgment, errors compound rapidly. When AI recommendations are routinely ignored because trust hasn't been established, the investment is wasted.

Clear ownership with explicit rules about when humans defer to AI and when they don't is essential.

Ethical and Social Judgment

AI models optimize for whatever they're trained to optimize for, without inherent ethical constraints. They can encode biases present in historical data. They can optimize in ways that are technically correct but socially unacceptable. They can make predictions that are accurate in aggregate but unfair to individuals.

Humans must provide the ethical framework governing AI use. What trade-offs are acceptable? What does fairness mean in context? What transparency or explainability is required? What outcomes are off-limits regardless of optimization potential?

An AI model might identify patterns predicting employee attrition with high accuracy, but using certain personal characteristics in these predictions may violate privacy principles or employment law. Human judgment determines what's appropriate, not technical capability.

How AI Fits Within Your Strategy Layer

The Strategy Layer provides the natural framework for deploying AI effectively. AI capabilities fit at specific points where they can accelerate or improve what's already defined.

From Data to Metrics: Automated Feature Extraction

Raw data becomes useful when transformed into meaningful metrics. AI can accelerate this transformation by automatically extracting relevant features from data, identifying leading indicators that predict important outcomes, or flagging anomalies requiring metric updates.

Example: A hospital system used AI to analyze electronic health record data and identify patterns predicting patient deterioration earlier than standard early warning scores. The AI didn't replace clinical judgment. It provided an additional metric (AI-predicted risk score) that clinicians incorporated into existing assessment processes.

The decision (whether to escalate care) remained with clinical staff. They had an additional, earlier indicator to inform that decision.

From Metrics to Decisions: Recommendations and Simulations

Once decision-critical metrics are defined, AI can help interpret them by generating recommendations about what to do or by simulating what different choices would produce.

AI processes complex combinations of metrics and recommends actions based on learned patterns of what worked in similar past situations.

Example: A manufacturing company used AI to recommend production schedules based on demand forecasts, capacity constraints, material availability, and quality history. The system generated recommended schedules that operations managers reviewed and approved.

Managers could override recommendations when they had information the AI didn't have (upcoming maintenance, quality concerns, strategic priorities). The AI made the first pass at a complex optimization problem. Humans made the final decision with context the AI lacked.

Continuous Monitoring and Early Warning

AI watches metrics continuously and alerts decision makers when patterns suggest upcoming issues, enabling earlier intervention. This is particularly valuable for high-frequency operational decisions where human monitoring would miss subtle early indicators.

Example: A financial services firm used AI to monitor transaction patterns for fraud risk. The AI system flagged suspicious patterns for investigation by fraud analysts. Over time, the system learned which patterns were genuinely concerning versus unusual but legitimate, improving precision and reducing false positives consuming analyst time.

The decision (whether to block a transaction or require additional verification) remained with analysts. AI directed their attention to highest-risk situations.

Scenario Planning Support

AI rapidly generates and evaluates scenarios, giving decision makers better understanding of options and trade-offs before committing.

Example: A retailer used AI to model scenarios for store staffing decisions: what customer experience, sales conversion, and labor cost outcomes would result from different staffing levels and schedules across various store formats and seasons.

Store managers used these scenarios to inform their staffing decisions, understanding trade-offs between cost and service more clearly than gut instinct alone would provide.

In each application, notice the pattern: AI augments human decision making within a structure the Strategy Layer provides. The decisions are defined, ownership is clear, metrics already exist or are being enhanced, and humans remain accountable for outcomes.

Risks of AI Without Strategy Layer Foundation

Organizations that deploy AI without the Strategy Layer foundation commonly experience predictable problems.

Automating Biased or Misaligned Decisions

AI trained on historical data replicates patterns in that data, including biases and misalignments.

If past hiring decisions were biased, AI trained on those decisions replicates the bias. If past resource allocation decisions optimized for local interests rather than strategic goals, AI automates that misalignment.

Without the Strategy Layer defining what decisions should optimize for and what constraints matter, AI becomes a high-speed mechanism for perpetuating whatever patterns existed historically, good and bad.

Example: A lending institution deployed AI to automate credit decisions, training models on historical approval patterns. The AI was technically accurate at predicting which applications historically would have been approved. But historical decisions had embedded biases the organization was trying to correct.

The AI automated and scaled those biases before anyone realized the problem. Untangling this required temporarily halting the AI system, carefully defining what fair lending decisions should look like (a Strategy Layer exercise), and retraining models with those criteria.

Over-Trusting Opaque Models

Many powerful AI techniques produce models that are difficult to interpret: neural networks, deep learning, ensemble methods. These "black box" models can be highly accurate but don't provide clear explanations for their predictions or recommendations.

Organizations without strong decision frameworks sometimes defer to these opaque models even when they don't understand why the model recommends what it does. This creates brittleness. When conditions change in ways the training data didn't include, models fail in ways humans don't recognize until damage is done.

The Strategy Layer provides scaffolding for appropriate AI trust: clear decision criteria that AI recommendations must satisfy, explicit thresholds where human override is required, and outcome monitoring revealing when AI performance degrades.

You can use black box models when you have transparent frameworks around them.

Generating Insights No One Acts On

AI can uncover fascinating patterns: customer segments you didn't know existed, leading indicators you hadn't considered, optimization opportunities you hadn't imagined.

These discoveries are only valuable if they connect to decisions someone can make.

Many AI initiatives produce dashboards full of ML-powered insights that executives find interesting but don't know how to act on. Without the Strategy Layer connecting insights to decisions with ownership and defined actions, AI becomes an expensive science project rather than a business capability.

Example: A retail chain deployed AI to analyze basket data and identify product affinities and customer microsegments. The models worked beautifully, revealing nuanced patterns in purchasing behavior.

But these insights didn't map to any decisions in the organization. Merchandising decisions were made at category level (too coarse for microsegments), pricing was set corporately (couldn't be customized to segments), and store operations couldn't act on product affinity insights.

The AI insights were accurate and interesting, and completely unused.

Guardrails for AI Deployment

Organizations that successfully deploy AI within their Strategy Layer follow key principles.

Only Apply AI Where Decisions Are Well-Defined

Don't start with AI. Start with Strategy Layer foundations: clear decisions, decision-critical metrics, established governance.

Once those exist, consider where AI could improve what's already working. AI accelerates good decision processes. It doesn't fix broken ones.

Test: Before deploying AI for a decision, verify that decision is documented in your inventory with clear ownership, defined metrics, and established review processes. If these foundations don't exist, build them first.

Maintain Human Accountability

AI can recommend, predict, simulate, or optimize, but humans decide and own outcomes.

Document who is accountable when AI is involved, under what conditions AI recommendations should be followed versus questioned, and how to escalate when AI and human judgment conflict.

Example: "AI-recommended credit decisions below $50K are approved automatically if fraud score is below threshold. Decisions above $50K or with elevated fraud scores require human review. Credit manager remains accountable for portfolio performance and can override AI recommendations with documented rationale."

Match Explainability to Decision Consequence

High-stakes decisions require more explainable AI. Humans need to understand why the AI recommends what it does.

Lower-stakes, reversible decisions can tolerate more opaque models if accuracy is higher.

Guideline: For decisions with irreversible consequences or significant ethical implications, require interpretable models or techniques that explain black box predictions. For routine operational decisions with limited individual impact, optimize for accuracy over explainability.

Monitor Outcomes Systematically

AI should improve decision outcomes, not just provide impressive predictions.

Track whether decisions informed by AI produce better results than previous approaches. When AI recommendations are overridden by human judgment, track whether humans or AI were right. Use this feedback to improve both AI models and human decision rules.

Example: "Monthly review compares outcomes when managers followed AI staffing recommendations versus when they overrode them. Analysis shows AI recommendations produce better customer satisfaction scores 70% of the time. We use override cases to identify situations where human judgment adds value (special events, known system issues, local market conditions) and incorporate those factors into future AI models."

Plan for Model Refresh and Retirement

AI models aren't permanent. They need retraining as conditions change, auditing for performance drift, and eventual retirement when they're no longer appropriate.

The Strategy Layer's lifecycle management (from Chapter 4) extends to AI models. They serve decisions, those decisions may change, and models should be retired when their decisions are deprecated.

Governance: Quarterly reviews of AI models in production assess whether they're still serving active decisions, whether performance meets standards, and whether underlying conditions have changed enough to require retraining.

Establish Clear Ethical Boundaries

Some outcomes are off-limits regardless of optimization potential.

Document these explicitly: constraints on what data can be used, what predictions are prohibited, what trade-offs are unacceptable, and what transparency or consent is required.

These guardrails prevent AI from technically succeeding while creating ethical or social problems.

Example: "AI models for hiring decisions may not use protected characteristics (race, gender, age) directly or proxy variables that correlate with protected characteristics. Models must achieve similar accuracy across demographic groups. Candidates have right to request human review of AI recommendations."

Practical Starting Points

Here's how to begin deploying AI within your Strategy Layer.

Start With One Well-Structured Decision

Choose one decision from your inventory where:

- The decision is made frequently (weekly or monthly)
- You have substantial historical data on decisions and outcomes
- Current decision process works but could be faster or more accurate
- Decision owner is willing to experiment with AI augmentation

Don't try to apply AI everywhere at once. Prove value in one focused application.

Build AI to Enhance Existing Metrics

Rather than creating entirely new AI-driven metrics, use AI to improve metrics you already track for this decision.

If you have a customer churn risk metric, enhance it with AI that detects earlier warning signals. If you have equipment failure predictions, improve them with AI that finds subtle leading indicators.

This approach integrates AI into existing decision processes rather than requiring process redesign.

Implement With Human-in-the-Loop

Deploy AI initially with human review of all recommendations. Don't automate immediately.

Let decision makers see AI recommendations alongside their current decision approach. Track when they agree, when they disagree, and what outcomes follow.

Build trust and understanding before increasing automation.

Phased approach:

- Phase 1 (3 months): AI generates recommendations, humans review all, data collected on agreement and outcomes
- Phase 2 (3 months): AI recommendations followed for low-stakes cases, human review for high-stakes, continued learning
- Phase 3 (ongoing): Selective automation where AI consistently outperforms humans, with periodic auditing

Measure Impact on Decision Quality

Define success by decision improvement, not AI accuracy alone.

Questions to answer:

- Are decisions made faster with AI support?
- Do decision makers have higher confidence?
- Are decision outcomes better (measured against actual results)?
- Where does AI add most value, and where does human judgment matter more?

If AI produces accurate predictions that don't improve decisions, it's not creating value regardless of technical sophistication.

A Realistic Example: Workforce Planning

Here's how AI deployment looks within a complete Strategy Layer for workforce planning decisions.

Decision from inventory: Monthly allocation of hiring budget across roles and locations

Decision owner: VP of Human Resources

Decision-critical metrics (from Chapter 4):

- Turnover rate by role and location
- Time-to-fill by role
- Cost of vacancy (revenue or productivity impact)
- Internal mobility rates
- External market salary data

Current process: HR reviews these metrics monthly, discusses with business unit leaders, allocates budget based on discussion and negotiation

AI enhancement approach:

Phase 1: Pattern detection Train AI model on 24 months of historical data relating turnover, hiring difficulty, business outcomes, and hiring decisions. Model identifies which factors most predict where hiring investment creates value versus where it's wasted.

Key insights: High-volume hiring in locations with high turnover and low engagement doesn't improve outcomes. Selective hiring targeting high performers in stable locations produces better business results. Certain roles have predictable seasonal turnover patterns allowing proactive hiring.

Phase 2: Prediction and recommendation AI generates monthly recommendations: "Based on current turnover trends, upcoming project needs, and historical success rates, recommend allocating 40% of budget to Location A (critical roles with high impact), 30% to Location B (stable area with growth opportunity), 20% to Location C (backfilling predictable attrition), 10% held in reserve for unexpected needs."

Recommendation includes rationale, key assumptions, and sensitivity analysis showing how results change under different scenarios.

Phase 3: Human review and decision VP of HR reviews AI recommendation in context of information AI doesn't have: upcoming reorganizations, strategic shifts, leadership changes, market conditions not yet in data.

VP accepts 80% of AI recommendations, modifies 15% based on strategic context, overrides 5% based on factors AI can't assess. Decision made and communicated within one day instead of two weeks of meetings.

Phase 4: Outcome tracking and learning Three months later, analysis shows:

- Locations following AI recommendations achieved hiring targets faster
- Roles where AI recommended higher investment showed better retention
- Cases where VP overrode AI mostly involved strategic context AI lacked

(appropriate overrides) but a few involved bias AI should learn to avoid

Learnings feed into next model iteration and into clearer documentation of when human judgment should override AI.

Result after one year:

- Hiring decision cycle time reduced from 2 weeks to 2 days
- Quality of hire improved (measured by 90-day retention and manager satisfaction)
- Budget efficiency improved (less wasted on roles/locations where hiring doesn't stick)
- VP confidence in decisions increased
- AI continues to learn from outcomes and human overrides

Notice how AI enhances but doesn't replace the Strategy Layer. The decision was already defined, owned, and supported by metrics. AI made the existing process faster and better, not fundamentally different.

Implementation Exercise: Assess AI Readiness for One Decision

Time required: 60 minutes

What you'll create: Clear assessment of whether a decision is ready for AI enhancement

Steps:

1. Select one decision (10 minutes)

Choose a decision from your inventory. Pick something that happens frequently and where you have historical data on decisions and outcomes.

2. Evaluate AI readiness (30 minutes)

Answer these questions:

Decision foundation:

- Is this decision well-defined in your inventory with clear ownership? (Yes/No)
- Do you have established metrics that inform this decision? (Yes/No)
- Is the decision made on a regular cadence? (Yes/No)

Data availability:

- Do you have at least 6-12 months of historical data on this decision? (Yes/No)
- Does historical data include both inputs (information considered) and outcomes (what resulted)? (Yes/No)
- Is the past likely to predict the future for this decision? (Yes/No)

Value potential:

- Would faster decisions create meaningful value? (Yes/No)
- Would more accurate predictions materially improve outcomes? (Yes/No)
- Can you specify exactly how you'd measure AI impact? (Yes/No)

Implementation feasibility:

- Is the decision owner willing to experiment with AI? (Yes/No)
- Can you implement with human-in-the-loop initially? (Yes/No)

- Do you have capacity to build, test, and maintain AI models? (Yes/No)

3. Score and interpret (10 minutes)

Count your "Yes" answers:

10-12 Yes: This decision is ready for AI enhancement. Proceed with pilot.

7-9 Yes: Decision has potential but needs some foundation work first. Address gaps before AI implementation.

4-6 Yes: Not yet ready. Strengthen Strategy Layer foundations (clearer decision definition, better metrics, more data collection) before considering AI.

0-3 Yes: Focus on basic decision support first. AI won't help until fundamental decision infrastructure exists.

4. Define next steps (10 minutes)

Based on your score:

If ready: Identify specific AI application (prediction, recommendation, optimization), data scientist or vendor to work with, success metrics, and timeline for pilot.

If not ready: List specific gaps to address (unclear ownership, missing metrics, inadequate data, etc.) and timeline to close them before reconsidering AI.

What you have now: Clear understanding of whether AI makes sense for this decision and what needs to happen before or during implementation.

Moving Forward

AI is powerful when deployed within clear decision frameworks. Without those frameworks, it's just sophisticated technology searching for problems to solve.

You've now learned how to:

- Build decision inventories and metrics (Chapters 3-4)
- Present decisions effectively and intake requests systematically (Chapters 5-6)
- Align cross-functional work around shared decisions (Chapter 7)
- Deploy AI to accelerate decision-making (Chapter 8)

The final chapters address how to scale these approaches across your organization and develop the leadership capabilities needed to make the Strategy Layer sustainable.

Chapter 9 shows you how to build Strategy Layer capability at enterprise scale, moving from initial success to organization-wide transformation.

You're ready to scale what you've learned.

Chapter 9: Scaling the Strategy Layer

You've built your first decision inventory. You've designed metrics that inform those decisions. You've created decision briefs that executives actually use. Your analytics intake now requires decision context before work begins. For one domain or strategic priority, the Strategy Layer is working.

Now comes the harder question: How do you scale this beyond your initial success? How do you move from one pocket of excellence to organization-wide capability?

This chapter provides the roadmap for building Strategy Layer capability at enterprise scale, moving from proof of concept to sustainable competitive advantage.

The Maturity Journey

Organizations evolve through recognizable stages as they build decision capability. Understanding these stages helps you assess where you are, what comes next, and what's required to progress.

Stage 1: Dashboard-Driven

Organizations at this stage have invested heavily in business intelligence: data warehouses, reporting tools, dashboards, visualization platforms. They can produce impressive-looking analytics.

Characteristics:

- Abundant dashboards covering many aspects of the business
- Metrics tracked because they seem important or someone requested them
- Analytics teams responsive to requests, building what's asked for
- Executives receive regular performance updates
- Decisions happen in meetings, often referencing data informally
- No clear connection between specific metrics and specific decisions

Pain points:

- Executives see lots of data but struggle to know what to do with it
- Analytics teams work hard but wonder if their work matters
- The same debates recur without resolution
- Metric definitions vary across teams, creating confusion

Most organizations spend years here, continuously expanding dashboards while remaining frustrated that better data doesn't produce better decisions. The problem isn't analytics quality. It's the absence of decision architecture.

Moving forward requires: Recognizing that dashboards and reports are tools, not strategies. They're valuable only when connected to clear decisions with ownership and accountability.

Stage 2: Decision-Aware but Fragmented

Organizations here have recognized that not all metrics are equally important. They've attempted to identify KPIs or critical success factors. Different parts of the organization may have adopted frameworks like OKRs or balanced scorecards.

Characteristics:

- Identified KPIs for different functions or business units
- Some attempt to connect metrics to strategic goals
- Executives and managers discuss metrics regularly
- Analytics organized by function, each supporting their domain
- Some decisions have established processes
- Growing awareness that metrics across functions sometimes conflict
- Pockets of excellence where decision support works well

Pain points:

- KPIs proliferate without clarity about priorities
- Cross-functional decisions remain difficult
- Strategy cascade loses coherence as goals fragment into functional interpretations
- Analytics teams duplicate effort or produce conflicting results
- Good practices don't transfer because there's no common framework

Organizations often plateau here, having made progress beyond generic dashboards but not achieving genuine integration.

Moving forward requires: Creating shared language and frameworks that span functions. This is exactly what the Strategy Layer provides.

Stage 3: Strategy Layer in Focused Domains

Organizations at this stage have implemented the Strategy Layer approach in specific domains where committed leaders championed it. These pockets demonstrate what's possible.

Characteristics:

- One or more domains (IT portfolio, workforce planning, customer experience) have functioning Strategy Layers
- Decision inventories exist for strategic priorities in these domains
- Metrics explicitly tied to decisions with clear ownership
- Decision review forums operate regularly
- Analytics work prioritized based on decision impact
- Visible success stories with proven ROI
- Growing recognition from leadership that this approach works
- Shared language spreading beyond initial pockets

Pain points:

- Success is localized; most of organization still operates in Stage 1 or 2
- Interfaces between mature and immature domains create friction
- Risk that champion departures could cause regression
- Difficulty scaling because each new domain feels like starting over
- Resource constraints as mature domains compete for analytics capacity

This stage is progress but not yet transformation. The organization has proven the Strategy Layer works but hasn't institutionalized it enterprise-wide.

Moving forward requires: Deliberate scaling strategies: capturing patterns from successful pockets, creating reusable frameworks, building organizational capability broadly, and securing sustained executive commitment.

Stage 4: Enterprise-Wide Capability

Organizations at this advanced stage have embedded the Strategy Layer as standard operating practice across the enterprise.

Characteristics:

- Decision inventories maintained for all strategic priorities
- Strategy Layer architecture (catalogs, metric maps, governance) established and current
- Analytics capacity allocated based on decision impact enterprise-wide
- Executives fluent in decision-focused language
- Cross-functional alignment through shared Strategy Layer frameworks
- Decision quality tracked at senior levels
- New initiatives automatically include Strategy Layer design
- Onboarding includes Strategy Layer orientation
- AI deployed within Strategy Layer guardrails
- Continuous improvement culture around decision quality

Capabilities:

- Faster, higher-confidence decision-making on critical choices

- Clear trade-off resolution when priorities conflict
- Analytics work focused on high-impact decisions
- Learning from decisions systematically captured and applied
- Strategy execution more reliable
- Competitive advantage from superior decision-making

Organizations at this stage don't have perfect decision-making. No organization does. But they have systematic, improvable approaches that create sustained advantage.

The Scaling Roadmap

Getting from Stage 1 or 2 to Stage 4 requires deliberate strategy, not organic spread.

Phase 1: Start Small and Prove Value

The most effective scaling begins with focused success rather than enterprise-wide mandate.

Choose one domain where several conditions align:

- Clear strategic importance (failure to improve would genuinely hurt)
- Leadership commitment (someone with authority will actively champion)
- Manageable scope (can implement in 3-6 months)
- Measurable outcomes (can demonstrate whether it's working)

Build the full Strategy Layer for that domain: identify critical decisions, create decision inventory, define decision-critical metrics, establish governance, connect analytics work to decisions, implement decision review processes.

Make this initial implementation work well. Invest time to get decision definitions right, metric definitions precise, and governance processes sustainable. Don't shortcut because you're eager to scale. The foundation you build here becomes the template for expansion.

Example: A national retailer started with store operations decisions: staffing allocation, inventory positioning, promotional calendars. These were strategically important, had clear ROI potential, and were contained enough to implement thoughtfully.

After six months, store operations had a functioning Strategy Layer with measurably better decisions, lower costs, and higher satisfaction from both store managers and analytics teams.

Phase 2: Capture Patterns and Create Templates

As your initial domain matures, deliberately capture what you've learned in reusable forms.

Decision templates by type: If you've built Strategy Layer support for investment decisions in one domain, create a template other domains can adapt for their investment decisions. Include common questions these decisions require, typical metrics that inform them, standard option frameworks, and proven governance patterns.

Metric definition patterns: Document how you defined decision-critical metrics: what made them work, what pitfalls you avoided, what governance processes maintain them. Create templates others can follow.

Governance approaches: Which review forums were valuable versus bureaucratic? What intake processes filtered requests effectively? How did you handle metric change control? Capture these as reproducible patterns.

Training materials: Build onboarding materials, workshop formats, and real examples from your initial domain that teach others how to implement the Strategy Layer without starting from scratch.

Measurement frameworks: Document how you assess Strategy Layer health in your domain. What indicators reveal whether decisions are being made well? These measurement approaches help other domains prove value as they implement.

This pattern capture shouldn't wait until everything is perfect. Start documenting lessons as soon as patterns become clear. Refine documentation as understanding deepens.

Phase 3: Extend to Additional Domains

With proven success and captured patterns, begin systematic extension. Prioritize based on:

Strategic importance: Which decisions matter most to organizational success? Where would improved decision quality create most competitive advantage?

Leadership readiness: Which leaders are eager to implement and have authority to drive change in their domains? Start with champions who will ensure success.

Interdependencies: Which domains have critical interfaces with already-mature areas? Sometimes the second domain is chosen because it shares decisions with the first, making alignment natural.

Learning value: Which domains would teach you most about scaling? Sometimes choosing a domain quite different from your initial one reveals whether patterns are genuinely reusable.

Expansion doesn't mean identical replication. Each domain adapts the Strategy Layer to its context: different decision types, different metrics, different governance cadences. But the core framework remains consistent.

Example: A healthcare system scaled from initial success in emergency operations to surgical services, then inpatient capacity management, then clinical quality, then workforce planning. Each domain took 4-6 months. Each learned from previous domains but adapted appropriately.

After three years, they had Strategy Layer coverage across all major operational domains, with enterprise-level coordination through shared frameworks.

Phase 4: Build Enterprise Infrastructure

As multiple domains mature, enterprise coordination becomes necessary and valuable.

Cross-domain decision mapping: Some decisions span multiple domains and require coordination. Create visibility into how domains connect and where alignment is needed.

Shared governance forums: Establish enterprise-level review of Strategy Layer health across domains. Where is it working? Where is it struggling? What cross-cutting issues need resolution?

Common architecture and tools: Standardize core components (decision catalog structure, metric definition templates, governance processes) while allowing domain-specific adaptation.

Analytics capacity management: As decision-focused prioritization becomes standard across domains, enterprise-level view of analytics capacity allocation becomes valuable. Are resources flowing to highest-impact decisions regardless of domain?

Enterprise learning systems: Create mechanisms for domains to share what they're learning about decision quality, metric design, and governance patterns. This accelerates improvement and prevents repeated mistakes.

Executive Strategy Layer oversight: Senior leadership explicitly reviews Strategy Layer maturity and impact as part of strategic planning and operational reviews.

This enterprise coordination shouldn't become bureaucratic overlay. It exists to enable domain success, not control or constrain. The right balance involves light-touch enterprise frameworks with significant domain autonomy in implementation.

Leadership Capabilities That Drive Scaling

Scaling the Strategy Layer requires leaders who can operate across levels that often remain disconnected: from strategic goals through specific decisions to metric design and data requirements, and back up to outcomes.

Decision Design Fluency

Leaders who scale the Strategy Layer can identify, structure, and document important decisions so they can be supported systematically.

They recognize when a decision exists versus when there's just a problem to monitor. They frame decisions clearly and specifically. They decompose large ambiguous decisions into manageable choices. They identify decision dependencies and sequences. They document decisions in ways others can understand and use.

Development approach: Own one decision end-to-end. Choose a decision you make regularly and apply the full Strategy Layer framework: document it clearly, identify questions it requires, define decision-critical metrics, establish review cadence, make the decision systematically, assess outcomes. Learn from this contained experience before trying to transform broader practices.

Metric Literacy

Leaders who scale the Strategy Layer can design, evaluate, and use metrics that actually inform decisions rather than just creating measurement.

They distinguish between vanity, health, and decision-critical metrics. They evaluate whether a metric is well-constructed. They connect metrics to decisions explicitly. They identify when metrics are misaligned across functions. They understand metric relationships and chains. They recognize when metrics are technically accurate but strategically misleading.

Development approach: Audit the top ten metrics your team uses against decision-critical criteria from Chapter 4. For each metric, ask: which decision does this inform? Does it have clear thresholds? Is there an accountable owner? Is it reviewed at decision cadence? Redesign metrics that don't pass this test.

Storytelling With Data for Influence

Leaders who scale the Strategy Layer can translate data into narratives that drive action, not just display facts.

They structure presentations around decisions and options, not just findings. They choose the right detail level for audience and purpose. They use visualization to clarify rather than decorate. They anticipate and address objections directly. They make recommendations with appropriate confidence. They connect data back to strategic outcomes throughout.

Development approach: Take one analysis you're planning to present. Before building slides, write out the decision brief structure from Chapter 5: context, decision, options, evidence, recommendation. Build your presentation from this structure. Notice how it changes what you emphasize and what you cut.

Systems Thinking and Governance

Leaders who scale the Strategy Layer can design processes that make decision quality sustainable, not just achieve one-time improvements.

They design governance that balances structure with flexibility. They build coalitions and secure executive sponsorship. They manage resistance constructively. They create shared language and mental models. They sustain attention over time through consistent reinforcement. They measure and communicate progress.

Development approach: Lead a Strategy Layer workshop for your team. Bring direct reports together for a half-day session. Work through the framework: identify strategic goals, catalog key decisions, map decisions to questions to metrics. Document the output and use it to guide your team's work. Notice what's hard about this facilitation and what makes it successful.

Sustaining Momentum

One of the greatest risks to Strategy Layer sustainability is leadership transition. When champions move on, will their successors continue the approach, or will it fade?

Embed in Operating Cadence

When Strategy Layer reviews are part of standard monthly or quarterly rhythms rather than special initiatives, they persist through leadership changes. New leaders inherit ongoing practices rather than having to rebuild.

Make decision review forums part of regular business reviews. Include decision quality metrics in performance dashboards. Reference the decision inventory in budget planning, strategic planning, and resource allocation processes.

Document Thoroughly

Clear documentation of decisions, metrics, and governance means new leaders can quickly understand what exists and why. They're not dependent on institutional memory that walks out the door with departing leaders.

Maintain accessible documentation: decision catalogs, metric definitions, governance processes, success stories, lessons learned. Make this documentation discoverable and usable, not buried in document management systems.

Demonstrate Results

When Strategy Layer approaches have measurably improved decision quality, speed, or outcomes, successors have incentive to continue rather than abandon them.

Track and communicate impact specifically: "Portfolio decisions now take 5 days instead of 3 weeks." "Customer retention improved 12% since implementing targeted decision support." "Analytics team satisfaction increased 30%."

Quantify the value so new leaders can see what they'd be giving up.

Build Broad Ownership

When Strategy Layer is owned by many leaders across domains rather than a single champion, it survives individual transitions.

If the CFO who drove Strategy Layer for financial decisions leaves but the COO, CMO, and CIO are all practicing it in their domains, the approach persists. Spread ownership intentionally during scaling.

Integrate Into Performance Expectations

When leaders are evaluated partly on decision quality in their domains, measured through Strategy Layer health metrics, successors understand this is a continuing expectation.

Include decision quality in leadership competency models. Ask about it in succession planning. Reference it in performance reviews. Make it part of "how we operate here."

Create Organizational Identity

When the organization describes itself as making decisions systematically based on clear strategy and solid evidence, this identity creates expectation that transcends individuals.

"This is how we operate here" is more durable than "this is what our previous leader preferred." Build this identity through consistent communication, celebration of success stories, and integration into onboarding.

Measuring Enterprise Progress

Track these indicators to know whether you're successfully scaling:

Coverage Metrics

Percentage of strategic goals with Strategy Layer support: Start tracking what percentage of your organization's strategic priorities have documented decision inventories, decision-critical metrics, and governance processes. Target 70%+ within 18-24 months.

Number of domains with mature Strategy Layers: Count how many organizational domains (functions, business units, strategic priorities) have moved beyond initial implementation to sustained operation. Track progression over time.

Percentage of analytics capacity aligned to decision inventory: What percentage of your analytics work supports decisions documented in inventories versus ad hoc requests? Target 60%+ for decision-aligned work.

Quality Metrics

Average decision cycle time for cataloged decisions: Are decisions getting made faster? Track time from recognizing need to decide until action happens. Target measurable reduction.

Decision maker confidence scores: Survey decision owners quarterly: "On a scale of 1-5, how confident are you that your critical decisions are well-informed?" Track the trend.

Percentage of decisions with post-decision outcome reviews: Are you systematically learning from decisions by reviewing whether they produced expected results? Target 80%+ of high-impact decisions.

Impact Metrics

Strategic goal achievement rate: Are you hitting more of your strategic goals since implementing Strategy Layer? This is the ultimate test.

Cross-functional conflict resolution time: When functions disagree about decisions, how quickly are conflicts resolved? Shared Strategy Layers should reduce resolution time significantly.

Analytics team satisfaction: Survey analysts quarterly about work meaningfulness. Do they feel their work influences important decisions? Rising satisfaction indicates successful Strategy Layer adoption.

A Realistic Timeline

Setting appropriate expectations about scaling pace prevents frustration and maintains commitment.

Months 1-6: Prove value in one domain Build complete Strategy Layer for initial domain. Demonstrate measurable improvement in decision quality or outcomes. Document lessons learned. Create initial templates.

Months 7-12: Extend to 2-3 additional domains Use patterns from initial success to accelerate implementation in new domains. Refine templates based on lessons learned. Begin building enterprise coordination mechanisms.

Months 13-18: Expand to 5-7 domains Scale continues with increasing efficiency as patterns mature and organizational capability builds. Address cross-domain coordination challenges. Establish enterprise governance.

Months 19-24: Enterprise infrastructure solidifies Majority of strategic priorities have Strategy Layer support. Enterprise coordination mechanisms functioning. Strategy Layer becomes "how we work" rather than "the new initiative."

Year 3+: Continuous improvement and extension Ongoing refinement of existing domains. Extension to new priorities as they emerge. Focus shifts from implementation to optimization and to extending to new parts of organization (acquisitions, new business units).

This timeline assumes dedicated effort and executive support. Organizations that treat Strategy Layer as something people do in addition to their regular jobs will progress more slowly.

Common Obstacles and Solutions

"We're too busy to add something new"

Reality check: You're not adding work. You're restructuring existing work to be more effective. Analytics teams are already building reports and dashboards. Decision makers are already making decisions. The Strategy Layer organizes this work around decisions rather than letting it happen reactively.

Solution: Start with one domain where pain is highest. Show that the new approach actually reduces wasted effort and produces better results. Use success to build momentum.

"This won't work in our culture"

Reality check: Culture isn't fixed. It evolves based on what behaviors get rewarded and what practices become standard. If your culture currently makes decisions through political negotiation rather than systematic frameworks, that's the problem the Strategy Layer solves, not a reason it can't work.

Solution: Find one executive champion willing to try it in their domain. Prove it works there. Culture shifts when people see better results and when new approaches become "how we do things."

"Different parts of our organization are too different"

Reality check: The Strategy Layer framework is flexible. The core concepts (decisions, metrics tied to decisions, governance) apply everywhere. Implementation details vary appropriately by context.

Solution: Embrace variation. Healthcare decisions look different from manufacturing decisions, which look different from financial services decisions. That's expected. The framework adapts while maintaining common language and principles.

"We don't have the analytics capacity"

Reality check: You have the capacity you have. The Strategy Layer helps you use it more effectively by focusing on decisions that matter rather than responding to whoever asks loudest.

Solution: Implement decision-required intake (Chapter 6). This alone often frees 20-30% of analytics capacity by eliminating low-value work. Use freed capacity for Strategy Layer implementation.

"Leadership keeps changing priorities"

Reality check: Priorities do change. The Strategy Layer makes this manageable by making decision structures explicit. When priorities shift, you update your decision inventory and associated metrics. The framework persists even as specific decisions evolve.

Solution: Build flexibility into your governance. Review decision inventories quarterly. Make updating them part of strategic planning. Frame the Strategy Layer as the operating system that adapts to changing strategy, not as rigid infrastructure that fights change.

Implementation Exercise: Build Your Scaling Plan

Time required: 2 hours

What you'll create: 12-month roadmap for scaling Strategy Layer in your organization

Steps:

1. Assess current state (20 minutes)

Where is your organization today?

- Stage 1 (Dashboard-driven): Still building analytics capability, no clear decision focus
- Stage 2 (Decision-aware but fragmented): Some decision thinking but not systematic

- Stage 3 (Strategy Layer in focused domains): Working in 1-2 areas, need to scale
- Stage 4 (Enterprise-wide): Already operating at scale, focus on optimization

2. Identify initial or next domain (20 minutes)

What domain should you focus on next?

List 3-5 candidates. For each, rate:

- Strategic importance (1-5)
- Leadership readiness (1-5)
- Implementation feasibility (1-5)
- Measurable impact potential (1-5)

Choose the highest-scoring domain as your next focus.

3. Define success metrics (15 minutes)

How will you know it's working in this domain?

Specify:

- Decision cycle time target (current state → future state)
- Decision maker confidence target (current → future)
- One outcome metric that should improve (specify baseline and target)
- Timeline for achieving these targets

4. Identify resources needed (20 minutes)

What do you need to succeed?

List:

- Executive sponsor for this domain
- Analytics capacity required (hours per week)
- Subject matter experts who need to participate
- Budget for tools, training, or external support (if any)
- Time commitment from decision owners

5. Create 12-month milestones (30 minutes)

Break the scaling plan into phases:

Months 1-3:

- Complete Strategy Layer for identified domain
- Document lessons learned and create templates
- Measure and communicate initial results

Months 4-6:

- Extend to 1-2 additional domains using templates
- Establish cross-domain coordination if needed
- Refine governance based on experience

Months 7-9:

- Expand to 2-3 more domains
- Build enterprise infrastructure as needed

- Scale training and capability building

Months 10-12:

- Consolidate gains across all implemented domains
- Assess maturity and identify gaps
- Plan Year 2 scaling strategy

6. Identify risks and mitigation (15 minutes)

What could derail this plan?

List top 3-5 risks and how you'll address them:

- Leadership transitions → Build ownership across multiple leaders
- Competing priorities → Secure executive sponsorship early
- Analytics capacity constraints → Implement intake discipline to free capacity
- Resistance from stakeholders → Start with willing partners, use success to convert skeptics
- Implementation fatigue → Pace extension appropriately, celebrate wins

What you have now: A concrete 12-month plan for scaling Strategy Layer with clear milestones, resource requirements, success metrics, and risk mitigation.

Next step: Review this plan with your executive sponsor. Adjust based on feedback. Commit to Month 1-3 deliverables and begin execution.

Moving Forward

You now have the complete Strategy Layer framework and the roadmap for scaling it:

Chapters 3-4: Core components (decisions and metrics) **Chapters 5-6:** Operational practices (briefs and intake) **Chapter 7:** Cross-functional alignment **Chapter 8:** AI deployment within frameworks **Chapter 9:** Enterprise scaling and capability building

The journey from dashboard-driven to Strategy Layer-enabled organization takes 18-36 months of sustained effort. But organizations that complete this journey create measurable, lasting advantages:

Decisions made faster with higher confidence. Strategy executed more reliably. Analytics work focused on what matters. Cross-functional conflicts resolved systematically. AI deployed responsibly within clear frameworks. Competitive advantage through superior decision-making that compounds over time.

The gap between strategy and execution closes not through better strategy documents or more sophisticated analytics, but through the deliberate architecture that connects them.

You have everything you need to build that architecture in your organization.

The only question is: will you start?

Appendix: Quick Reference Guides

Decision Inventory Quick Start

1. Choose one strategic goal

2. List 10-15 decisions that drive it

3. Prioritize to top 3-5

4. Document each: name, owner, frequency, current state, strategic link

5. Validate with decision owners

Metric Design Quick Start

1. Start with documented decision

2. List 3-5 essential questions

3. Design 1-2 metrics per question

4. Specify: definition, calculation, sources, thresholds

5. Validate with decision owner

Decision Brief Template

- **Context:** What's at stake (2-3 sentences)
- **Decision:** What needs to be decided (clear question)
- **Options:** 2-4 alternatives with expected outcomes
- **Evidence:** 3-5 decision-critical metrics with interpretation
- **Recommendation:** Clear position with rationale

Analytics Intake Template

- **Decision to be made:** [Specific choice]
- **Decision owner:** [Name/role]
- **Decision deadline:** [Date]
- **Options under consideration:** [Alternatives]
- **What changes based on results:** [Actions]

Cross-Functional Alignment Process

1. Convene stakeholders around one outcome
2. Map shared decisions
3. Define questions for each decision
4. Identify metrics from each function
5. Establish thresholds and trade-off rules
6. Create shared review forum

AI Readiness Assessment

- Is decision well-defined with clear ownership?
- Do you have established metrics?
- Is there 6+ months historical data?
- Would faster/better predictions create value?
- Is decision owner willing to experiment?
- Can you implement human-in-the-loop?

If mostly yes: Proceed with pilot. If mostly no: Strengthen foundations first

Scaling Checklist

☐ Proven success in one domain ☐ Patterns captured in templates ☐ Next 2-3 domains identified ☐ Executive sponsors committed ☐ Analytics capacity allocated ☐ Training materials created ☐ Enterprise coordination planned ☐ Success metrics defined ☐ 12-month roadmap documented ☐ Risk mitigation strategies in place

The Strategy Layer transforms how organizations connect strategy to execution. You've learned the framework. You have the tools. The only thing remaining is implementation.

Start with one decision. Prove it works. Then scale.

Your organization's competitive advantage depends on how well you make decisions. The Strategy Layer is how you make decision-making a designed capability rather than an emergent behavior.

Begin building yours today.

Appendix A: Quick Reference Guides

Decision Inventory Quick Start

1. **Choose one strategic goal**
2. **List 10-15 decisions that drive it**
3. **Prioritize to top 3-5** based on:
    - Strategic impact
    - Decision frequency
    - Current quality gap
    - Feasibility
4. **Document each decision:**
    - Decision name and description
    - Owner
    - Frequency and timing
    - Current state, pain points

    - Strategic link

5. **Validate with decision owners**

Metric Design Quick Start

1. **Start with documented decision**
2. **List 3-5 essential questions** that decision requires answered
3. **Design 1-2 metrics per question**
4. **Specify for each metric:**
    - Definition (precise specification)
    - Calculation method
    - Data sources
    - Update frequency
    - Ownership
    - Thresholds that trigger action
5. **Validate with decision owner**

Decision Brief Template

Context (2-3 sentences):

- What's at stake?
- Why does this need to be decided now?

Decision (1-2 sentences):

- State the explicit choice being made
- Clarify what's in scope and what's not

Options (2-4 alternatives):

For each option:

- Description: What would we actually do?
- Expected outcome: What results do we anticipate?
- Key trade-offs: What are we giving up or risking?
- Timeline and cost: When and how much?

Evidence (3-5 decision-critical metrics):

For each key metric:

- Current value and trend
- What this tells us about options
- Data confidence level

Recommendation:

- Which option do you recommend?
- Primary rationale
- Key risks acknowledged
- Conditions that might change recommendation

Analytics Intake Template

Decision Context (Required):

- Decision to be made: [Specific choice]
- Decision owner: [Name/role]
- Decision deadline: [Date]
- Options under consideration: [Alternatives]
- What changes based on results: [Actions]

Current Situation:

- What prompted this request:
- What you already know:
- What's unclear or unknown:

Strategic Connection:

- Which strategic goal does this influence:
- Estimated impact if decided well:
- Consequence if decided poorly:

Cross-Functional Alignment Process

1. **Convene stakeholders** around one strategic outcome
2. **Map shared decisions** that drive that outcome
3. **Define questions** for each decision (from each function's perspective)

4. **Identify metrics** each function can provide

5. **Establish thresholds and trade-off rules**

6. **Create shared review forum** (regular cadence)

AI Readiness Assessment

Check these criteria:

☐ Decision is well-defined with clear ownership ☐ Established metrics already exist ☐ 6+ months of historical decision data available ☐ Faster/better predictions would create measurable value ☐ Decision owner willing to experiment with AI ☐ Can implement human-in-the-loop initially ☐ Have capacity to build, test, and maintain models

If mostly yes: Proceed with pilot **If mostly no:** Strengthen Strategy Layer foundations first

Scaling Checklist

☐ Proven success in one domain ☐ Patterns captured in reusable templates ☐ Next 2-3 domains identified ☐ Executive sponsors committed ☐ Analytics capacity allocated ☐ Training materials created ☐ Enterprise coordination planned ☐ Success metrics defined ☐ 12-month roadmap documented ☐ Risk mitigation strategies in place

APPENDIX B: COMPLETE TEMPLATES

Template 1: Decision Inventory Document

Decision Name: [Clear statement of what's being decided]

Owner: [Role or person with final authority]

Frequency: [How often this occurs and typical timing]

Description: [2-3 sentences explaining the choice and why it matters]

Current Process: [Brief description of how this is done today and key pain points]

Stakeholders: [Who provides input or needs to be consulted]

Strategic Link: [Which strategic goal this influences and how]

Priority: [High/Medium/Low based on impact, frequency, and quality gap]

Maturity Assessment: [Current state: Ad hoc / Defined / Managed / Optimized]

Decision-Critical Metrics:

1. [Metric name and definition]
2. [Metric name and definition]
3. [Metric name and definition]

Next Actions: [Immediate steps to improve this decision's support]

Template 2: Metric Definition Standard

Metric Name: [Clear, descriptive name]

Purpose: [Which specific decision this informs and how]

Definition: [Precise specification of what's being measured]

Calculation Method: [Exactly how the metric is computed]

Formula: [If applicable: mathematical formula]

Data Sources:

- Primary: [System or process providing the data]
- Secondary: [Additional sources if needed]
- Update schedule: [How often source data refreshes]

Update Frequency: [How often calculated and reviewed]

Owner:

- Calculation: [Who's responsible for maintaining the metric]
- Action: [Who's accountable for acting on results]

Thresholds:

- [Level 1]: [Value] means [interpretation] triggers [action]
- [Level 2]: [Value] means [interpretation] triggers [action]
- [Level 3]: [Value] means [interpretation] triggers [action]

Validation:

- Last reviewed: [Date]
- Reviewed by: [Decision owner confirmation]
- Next review: [Date]

Related Metrics: [Other metrics that provide context]

Change History:

- [Date]: [Description of change and rationale]

Template 3: Decision Brief (One-Page Format)

DECISION BRIEF

Prepared for: [Decision owner and stakeholders]**Prepared by:** [Your name and role]**Date:** [Date prepared]

CONTEXT [2-3 sentences: What's at stake? Why now?]

DECISION [1-2 sentences: State the explicit choice being made]

OPTIONS

Option A: [Name]

- Actions: [What we would do]
- Outcomes: [Expected results]
- Trade-offs: [What we're giving up/risking]
- Investment: [Cost and timing]

Option B: [Name]

- Actions: [What we would do]
- Outcomes: [Expected results]
- Trade-offs: [What we're giving up/risking]
- Investment: [Cost and timing]

EVIDENCE

RECOMMENDATION

We recommend **Option [A/B]** because [primary rationale].

Key risks: [Acknowledge main risks]

Template 4: Analytics Request Form

ANALYTICS REQUEST

Requester Information

- Name:
- Department:

- Email:
- Date submitted:

Work Type (check one) □ Decision support (inform specific decision) □ Exploratory analysis (hypothesis development) □ Infrastructure (data quality, systems) □ Self-service enablement (building tools)

For Decision Support Requests:

Decision Context

- Decision to be made:
- Decision owner:
- Decision deadline:
- Options being considered:
- What changes based on results:

Current Situation

- What prompted this request:
- What you already know:
- What's unclear:

Desired Outcome

- Specific questions needing answers:
- How results should be presented:
- Who else needs to see this:

Strategic Connection

- Strategic goal this influences:

- Impact if decided well:
- Consequence if decided poorly:

Priority Self-Assessment ☐ High (decision in next 2 weeks with major impact) ☐ Medium (decision in next month with moderate impact) ☐ Low (timing flexible or impact limited)

Template 5: Decision Review Meeting Agenda

DECISION REVIEW: [Decision Name]

Date: [Meeting date]**Attendees:** [Decision owner, key stakeholders]**Duration:** 60 minutes

1. Decision Performance Review (15 min)

- Decisions made since last review: [Number]
- Average time from recognition to action: [Days]
- Decision maker confidence (1-5 scale): [Score]
- Outcome assessment: [% achieving intended results]

2. Metric Review (20 min) For each decision-critical metric:

- Current value and trend
- Alerts or threshold triggers
- Data quality issues
- Need for metric adjustments

3. Process Improvement (15 min)

- What's working well in current decision process
- Where stakeholders are frustrated

- Information gaps affecting decision quality
- Recommendations for process changes

4. Strategic Alignment Check (10 min)

- Does this decision still serve current strategic goals?
- Any changes in priorities affecting this decision?
- Resource allocation appropriate?

5. Action Items and Next Steps (10 min)

- Decisions or commitments from this review
- Owner and deadline for each action
- Date for next review

Template 6: Shared Decision Framework (Cross-Functional)

SHARED STRATEGY LAYER

Strategic Outcome: [The goal requiring cross-functional collaboration]

Critical Decisions:

Decision 1: [Name]

- Owner: [Role]
- Frequency: [Timing]
- Description: [What's being decided]

Essential Questions: From [Function A]:

- [Question requiring their expertise] From [Function B]:

- [Question requiring their expertise] From [Function C]:
- [Question requiring their expertise]

Decision-Critical Metrics:

- [Function A] provides: [Metric name and definition]
- [Function B] provides: [Metric name and definition]
- [Function C] provides: [Metric name and definition]

Thresholds:

- When [metric] reaches [value], [action] is triggered
- When [metric] reaches [value], [action] is triggered

Trade-off Rules:

- When [metric A] and [metric B] conflict, prioritize based on: [rule]
- When [metric C] shows [condition], override other considerations

Escalation Path:

- If functions cannot reach agreement within [timeframe]
- Escalate to: [Role]
- With: [Required information package]
- Timeline for resolution: [Duration]

Shared Review Forum:

- Frequency: [Cadence]
- Attendees: [Roles from each function]
- Agenda focus: [Key review points]

Appendix C: Extended Case Studies

Case Study 1: Healthcare System - Patient Throughput

Organization: Regional health system with 4 hospitals, 800 beds, 3,500 staff

Challenge: Patient throughput declining, ED wait times increasing, bed utilization inconsistent across facilities, no coordinated capacity management

Strategy Layer Implementation:

Phase 1: Decision Inventory (Month 1-2) Identified 5 critical decisions:

1. Daily bed allocation across services
2. Surge protocol activation timing
3. Surgical scheduling prioritization
4. Discharge planning resource allocation
5. Inter-facility patient transfer

Phase 2: Metric Design (Month 2-3) Created decision-critical metrics:

- Current and predicted census by service (updated hourly)

- Discharge readiness status by patient (updated daily)
- ED pipeline forecast (updated every 4 hours)
- Nurse staffing ratios by unit (real-time)
- Surgical case duration predictions (updated with each case)

Phase 3: Decision Brief Implementation (Month 3-4)

- Replaced 90-minute capacity meetings with 30-minute decision reviews
- Standardized bed allocation decision brief reviewed by operations at 7am daily
- Clear criteria for when to activate surge protocols

Phase 4: Cross-Functional Alignment (Month 4-6)

- Brought together ED, inpatient units, surgical services, nursing, and capacity planning
- Established shared trade-off rules (safety > emergency admissions > scheduled surgery > elective admissions)
- Created daily capacity coordination meeting using shared dashboard

Results After 12 Months:

- ED wait times reduced from 87 minutes to 52 minutes (40% improvement)
- Bed utilization improved from 73% to 85% while maintaining quality
- Surgical case cancellations reduced by 60%
- Staff satisfaction with capacity decisions increased from 2.3 to 4.1 (5-point scale)
- Estimated revenue impact: $12M annually from improved throughput

Key Success Factors:

- Executive sponsor (COO) maintained consistent focus
- Started with visible pain point (ED wait times)
- Proved value quickly (3-month pilot showed 25% improvement)
- Built on existing data (didn't require new systems)
- Made trade-offs explicit rather than negotiated

Lessons Learned:

- Initial resistance from surgical services (felt bed allocation limited their autonomy)
- Required 6 weeks of parallel operation (old and new process) to build trust
- Metric definitions needed 3 iterations to get stakeholder agreement
- Cultural shift took longer than technical implementation

Case Study 2: Manufacturing Company - Portfolio Investment

Organization: Global manufacturer, $2B revenue, 800 active projects, distributed decision-making across 6 business units

Challenge: Project portfolio disconnected from strategy, unclear funding criteria, projects approved but under-resourced, no systematic review of outcomes

Strategy Layer Implementation:

Phase 1: Decision Inventory (Month 1) Mapped 3 critical portfolio decisions:

1. Quarterly funding allocation across project types
2. Monthly resource reallocation (people, budget)

3. Project cancellation or deferral triggers

Phase 2: Metric Design (Month 1-2) Defined decision-critical metrics:

- Strategic alignment score (1-5 scale, executive-rated quarterly)
- Schedule variance percentage by project
- Technical debt accumulation index
- Resource utilization by skill type
- Expected vs. actual business value delivery

Phase 3: Decision Brief Format (Month 2-3) Created standard portfolio review brief:

- Investment options (continue current portfolio vs. 3 rebalancing scenarios)
- Evidence (alignment scores, schedule variance, resource constraints, value delivery)
- Trade-off analysis (strategic value vs. delivery confidence vs. resource availability)
- Clear recommendations with rationale

Phase 4: Governance (Month 3-6)

- Established quarterly portfolio reviews using decision brief format
- Required all new project requests to specify which decision they relate to
- Created monthly resource reallocation meetings with clear criteria

Results After 18 Months:

- Portfolio alignment score improved from 2.8 to 4.2 (strategic fit)

- Project completion rate increased from 62% to 81%
- Resource contention issues (escalations) decreased 70%
- Business value delivery improved 35% (measured 12 months post-completion)
- Decision cycle time reduced from 6 weeks to 1 week for portfolio changes

Scaling: After initial success with IT portfolio, expanded to:

- Product development portfolio (Month 12)
- Capital investment decisions (Month 15)
- M&A evaluation framework (Month 18)

Key Success Factors:

- CFO and CTO co-sponsored (finance and technology alignment)
- Tied to annual planning cycle (natural review point)
- Used existing project data (minimal new collection)
- Made trade-offs visible (technical debt vs. new features)

Challenges Overcome:

- Business unit leaders initially resisted centralized framework
- Solution: Kept decision authority distributed but standardized criteria
- Project managers concerned about more governance overhead
- Solution: Replaced 3 separate review processes with 1 integrated review
- Debate about strategic alignment scoring methodology
- Solution: Executive team workshop to calibrate scoring together

Case Study 3: Financial Services - Customer Retention

Organization: Regional bank, 250K customers, 40 branches, losing customers to digital-first competitors

Challenge: Customer churn increasing, retention efforts reactive and inconsistent, no clear ownership of retention strategy, multiple conflicting initiatives

Strategy Layer Implementation:

Phase 1: Decision Inventory and Alignment (Month 1-2) Cross-functional team (Sales, Service, Product, Marketing, Analytics) identified decisions:

1. Which at-risk customers to target for retention programs
2. What retention tactics to deploy by customer segment
3. How to allocate retention budget across channels
4. When to escalate at-risk situations to relationship managers

Phase 2: Shared Metrics (Month 2-3) Functions contributed complementary metrics:

- Sales: Customer relationship depth score
- Service: Issue resolution time and satisfaction
- Product: Product usage patterns and engagement
- Marketing: Campaign response rates by segment
- Analytics: Churn risk prediction model (0-100 score)

Phase 3: Decision Framework (Month 3-4) Established shared decision rules:

- Churn risk >70 + relationship value >$50K = relationship manager contact within 48 hours
- Churn risk 50-70 = automated retention campaign

- Churn risk <50 = standard engagement
- Budget allocation: 60% to high-value accounts, 30% to growth segments, 10% experimental

Phase 4: Review Cadence (Month 4-6)

- Weekly retention operations review (30 minutes, cross-functional)
- Monthly program effectiveness analysis
- Quarterly strategic alignment and budget review

Results After 12 Months:

- Customer churn reduced from 12.3% to 8.7% annually
- Retention program ROI: $4.8M saved revenue vs. $1.2M program cost
- Customer lifetime value increased 18% for retained high-value customers
- Time from churn risk detection to intervention reduced from 12 days to 2 days
- Cross-functional conflict resolution time reduced 85%

Analytics Transformation:

- Analytics requests requiring decision context increased from 15% to 82%
- Rework rate for analytics projects decreased from 45% to 8%
- Analyst satisfaction with work meaningfulness increased from 2.9 to 4.4

Scaling Beyond Retention: Used same approach for:

- Credit risk decision framework (Month 9)
- Branch network optimization decisions (Month 12)

- Digital product investment priorities (Month 15)

Key Success Factors:

- Clear executive sponsor (Chief Customer Officer)
- Quick win focus (prevented 200 high-value customer churns in first quarter)
- Leveraged existing churn prediction model (added decision framework around it)
- Made budget allocation decisions explicit and rules-based

Cultural Shift:

- From: "We need more customer data" → To: "What decision does this data support?"
- From: Function-specific metrics and goals → To: Shared metrics for shared outcomes
- From: Meeting to discuss churn → To: Meeting to decide on interventions
- From: Attribution battles → To: Joint accountability for retention rate

About the Author

Steven understands your frustration.

You've invested in analytics. You have talented people, solid data, impressive dashboards. Yet when it's time to make important decisions, something's still missing. The connection between all that information and the choices you need to make remains frustratingly unclear.

Steven has spent his career working in that exact space. Here's what he's learned: the problem isn't your data, your tools, or your people. It's that no one ever taught us how to build the bridge between strategy and execution. We assumed good data would naturally lead to good decisions. It doesn't.

He's sat in meetings where smart executives reviewed polished dashboards and still decided based on gut instinct. He's worked with analytics teams who built brilliant analysis that no one used. Through trial and error, he figured out what actually works.

Steven has influenced over $100 million in investment decisions and built decision frameworks for hundreds of enterprise initiatives across healthcare, biotechnology, and operations. But his real expertise isn't in any particular sector. It's understanding how organizations actually make decisions and what it takes to make them better.

He's made his share of mistakes. Dashboards executives looked at once and never opened again. Metrics that seemed important but no one acted on. He learned that more data doesn't equal better decisions, that visibility doesn't equal impact, and that what looks good in theory often falls apart in practice.

The Strategy Layer is what he's learned actually works. Not in ideal conditions, but in the real world where you have limited capacity, competing priorities, messy data, and decisions that can't wait for perfect information.

If you've wondered why your analytics investments haven't made decisions easier, or you're tired of watching good analysis fail to influence important choices, this book is for you. A practical guide from someone who's been in the trenches and figured out what it takes to close the gap.

Made in the USA
Monee, IL
04 November 2026

67382717R00115